Marcio Luis Fernandes

A Place in Rio

Marcio Luis Fernandes

A Place in Rio

About the geographies of Ilha de Guaratiba

ScienciaScripts

Cover image: www.ingimage.com

This book is a translation from the original published under ISBN 978-3-330-75717-2.

Publisher:
Sciencia Scripts
is a trademark of
Dodo Books Indian Ocean Ltd. and OmniScriptum S.R.L publishing group

120 High Road, East Finchley, London, N2 9ED, United Kingdom
Str. Armeneasca 28/1, office 1, Chisinau MD-2012, Republic of Moldova, Europe
Printed at: see last page
ISBN: 978-620-8-28644-6

SUMMARY

A place in Rio

Marcio Luis Fernandes

DEDICATORY

I consecrate this work to my Good God, the source of all my inspiration. He is the Lord of everything and everyone, the master of all things. He sent his only son, Jesus Christ, Lord and finisher of my faith. He has been with me at all times, not allowing this thesis to become a solitary work. "For from Him and through Him and to Him are all things" (Romans 11: 36).

I also offer this research to my dear mother, Vandi Bastos de Moraes. Without the support of this incredible woman, I certainly wouldn't have made it this far. She has always been my greatest supporter and for this reason she deserves my affection, admiration and love.

I dedicate this work to the greatest gift I have ever received from heaven. I'm talking about my daughter Nicole Bazzani Fernandes, who was born, raised and developed during the course of this thesis. My life took on a much broader meaning after her.

I would also like to dedicate this research to my dear wife Jeane Bazzani, my partner for all hours. My days would be much less interesting without your company! Thank you for your understanding and support at all times.

I couldn't help but pay tribute to my dear neighbors and friends from Guaratiba. In fact, this investigative work was written because of the residents of Ilha de Guaratiba. They are the true authors of this book.

ACKNOWLEDGMENTS

I owe a huge debt of gratitude to my advisor, Professor Joâo Baptista Ferreira de Mello. It was certainly he who formed me as a researcher. I met him during the first period of my degree, in 2001, through a text that inspired me, from the outset, to adopt the humanistic approach to geography (MELLO, 1991). In fact, I came to UERJ because of you. In the specialization course in Territorial Policies in the State of Rio de Janeiro, my goal was to get to know better and get closer to the professor who was famous for his geographical routes through the Marvellous City, as well as for his unprecedented approach to geography and music. I'm glad it worked out. From 2005 to 2015, it was a total of nine years of mentoring and partnership (specialization: 2005-2006; master's degree: 2008-2009-2010; doctorate: 2012-2013-2014-2015). Thanks for everything Joâo...

Another person who made a big difference in my life was Professor Miguel Ângelo Campos Ribeiro. Before I met Miguel, his fame had already spread throughout the campus of the Moacyr Sreder Bastos University Center. The presence of one of the great stars of Brazilian geography at our faculty was always a source of honor, with the auditorium always packed with undergraduates willing to listen to his creative approaches to diverse topics such as the interaction between tourism and geography, the territories of prostitution, reflections on the space of Rio de Janeiro, among others. Since my arrival at UERJ in 2005, Miguel's participation in my university career has always been emblematic. I have always been impressed by his teaching and his charisma in the classroom. He was present at all my exams at UERJ (specialization monograph, master's degree qualification, dissertation defence, doctoral qualification), always contributing to my improvement. My sincere thanks go to my dear external Professor Miguel.

I was very happy to be able to count on the participation of Professor Aureanice de Mello Corrêa on the panel for my doctorate qualification. Her comments and suggestions contributed greatly to the development of this text. Professor Aureanice is certainly the friendliest and most outgoing professor in the

Geography Postgraduate Program at Rio de Janeiro State University. When she participates in UERJ events, she always makes the atmosphere more relaxed with her light-hearted presentations that escape conventional academic rigor. Her laughter always brings joy to the halls and corridors of our university. The research carried out at NEPEC (Nùcleo de Estudos e Pesquisas sobre Espaço e Cultura), studies related to the cultural dimension of space, linked to the religious phenomenon, gave Professor Aureanice know-how in relation to interpretative studies. I am grateful for her contribution, for her selflessness and for further brightening the defense of her doctorate in geography.

I would also like to thank Professor Lucia Maria Sà Antunes Costa, a full professor at the Federal University of Rio de Janeiro and coordinator of the Professional Master's Degree in Landscape Architecture at PROURB-FAU/UFRJ, for the privilege of being able to count on her presence at this thesis defense. Following in the footsteps of the late Roberto Burle Marx and in common with many other residents of Ilha de Guaratiba, whose stories can be found in this research, Professor Lùcia Costa is also a landscape architect. Her keen eye for the industry in evidence on Ilha de Guaratiba will certainly contribute to enriching the debates to which I will be subjected during the interview. Thank you for being willing to contribute and participate in my defense.

Professor Werther Holzer, a professor in the Postgraduate Program in Architecture and Urbanism at the Fluminense Federal University (PPGAU/UFF), is considered, along with my advisor, to be the greatest exponent of humanistic geography (or humanist geography, as he prefers) in Brazil, the forerunner of this branch of geographical knowledge in our country. Between the 1980s and 1990s, a time when the currents of geographic thought that prevailed in Brazil still disregarded the relationships of individuals and social groups with regard to their respective spaces and places - based on Tuan, Lowenthal, Buttimer, among others - Werther Holzer brought to light the philosophical presuppositions of humanism in geography. From then on, a new possibility arose in Brazilian geography for approaches to reality centered on human beings in the midst of their lived universe. Men and women are no longer considered just another element of the landscape, counted in droves like cattle. I feel extremely honored to have the two geographers considered to be the managers of Brazilian humanistic geography at my doctoral thesis defense, namely Professors Joào Baptista Ferreira de Mello and Werther Holzer.

Like a bird that wanders far from its nest, so is **man** who wanders far from his **place**.

(Proverbs 27: 8)

FIRST WORDS

My childhood was spent among guava trees, orange trees and banana trees in a very Brazilian and Rio de Janeiro style, but with the distinctive touches of a peri-urban suburb. Games included flying kites, pique-flag races, pelada and marbles. As my father, Manoel Fernandes, was a gravedigger, I was obliged to take his lunch to Ilha de Guaratiba Cemetery. That's why I had no feelings of rejection for the field of the dead. As a teenager, because I had few resources, I started working as a market vendor selling vegetables that I had grown myself. All this led me to an experience of strong attachment to my lived universe and its people, internalized within me.

In this intellectual journey, intertwined with my place, I use my lived experience and, together with the members of my community, I undertake this saga about our lived universe. This memorable place, inscribed in our hearts and minds, Ilha de Guaratiba, has been metamorphosed by ambition and speculative perspectives on its domains. I was born here. I saw the neighborhood with its rural/agricultural physiognomy, in previous decades, now replaced more strongly by urban residential with a view to constituting a spatial portion of enormous economic value of its soil, on behalf of real estate agents counting on the incentive/ganance of the Public Power. More recently, we have witnessed what could be called an Olympic river flowing over Ilha de Guaratiba, which is involved in its meanders, also flowing through its channel.

In fact, there are as many geographies as there are interpretations of the lived world of each individual and/or social group (JOHNSTON, 1986 a). Based on the idea that "knowledge of place is grounded in experience" (HOLZER, 1992), humanistic geographers have focused on lived experiences, looking to the stories of ordinary people as the foundation for their research. In this approach, the human being is the geographic actor who, in transforming their environment, also modifies their life relationships (JOHNSTON, 1986 b). This is because, in the phenomenological inseparability that guides humanism in geography, the individual is not distinct from their place (RELPH, 1976). Therefore, when a place undergoes transformations of various kinds, or those caused by unhappiness, existential changes can occur. In these cases, it is also common for people's relationship with their lived world to change (FERNANDES, 2003; 2006; 2010).

Pursuing a break with the positivist and neopositivist paradigms, whose approaches favored only the tangible and quantifiable aspects of reality, phenomenology, from the outset, pushed supporters of humanism in geography to shift their attention from spatial systems to the human experience in its meanings, whether negative or positive (BUTTIMER, 1982; MELLO, 2005). Genuine knowledge of place, however, involves both its subjective values: affective experiences, intimate experiences, belonging, disappointments, symbolism, meanings, the life of relationships and the existential way of life of its inhabitants - and its objective characteristics and values: physical particularities, internal structure, dynamic dimension, natural attributes, location, real estate value, level of urbanization, etc. (LOWENTHAL, 1982; FERNANDES; 2009; 2012). It is worth emphasizing that these dimensions are inseparable, considering the humanistic approach in its phenomenological stance, which privileges the inseparability of subject/object.

By assuming that knowledge does not exist independently of the human being, phenomenology claims that geographical knowledge needs to be obtained through the human experience of the world. In this sense, the phenomenological method is offered to geographers as a procedure in their approaches to place (RELPH, 1970). In their investigations, lived experiences are fundamental to unveiling the multidimensionality of place, endowed with "physical features, observable activities and functions, and meanings or symbols" (HOLZER, 1992, P. 251).

Because they are visible, observable facts and phenomena can "serve as a bridge between the objective and subjective approaches to geography" (SACK, 1976). In its classic conception, humanism values existence over objects and/or the material world (HOLZER, 1992). The main concern of humanistic geography has always been related to the experience of individuals in their daily activities (BUTTIMER, 1982; MELLO, 2005). What is lived, however, presupposes the material and the non-material; the objective and the subjective; the place in its morphology and spatial dynamics and its experiencers in their meanings and experiences (FRÉMONT, 1982). How can we neglect materiality and the spatial events that jump out at us? Doesn't the material world around us, as well as the dynamics we are surrounded by, influence our existence?

In the words of Tuan (1961, P. 32), "we are under no obligation to describe any area other than that for which we have a special affection or an inexplicable fascination". The place in which we live is our fundamental focus of concern. Our world is not derived from scientific studies, but from our experiences of it (PICKLES, 1985). In accordance with these precepts, the book is based on the principles of humanism in geography, and is developed in the context of the author's own place, based on his own experiences, as well as the experiences of the members of his community.

In these circumstances, the affective link between the person and the place, forged through lived experiences, mediated - most of the time - by a long and/or intense interaction between the individual and their lived world, leaps out to be elucidated in the midst of academic canons. These values are singular, collective, subjective and intersubjective, thus forming part of the intimate and private collections of individuals and social groups. In his humanistic approach to the relationship between space and place, Mello (1990), based on Tuan (2013), asserts that certain spaces only become places after a long experience. What is initially ugly, dull or even hated (space) can, over time, become a place. Spaces become places because of contact with other people and because of emotional, economic and other exchanges (MELLO, 1990). In this context, spaces become places not only through affective exchanges (valuation), but also through economic exchanges (valuation) or for reasons of status (YAZIGI, 2003). Certain spaces only become places after they have undergone a process of valorization that brings about a change in those who experience them (valorization) in their relationship with their daily lives and/or lived world. In this case, the valorization of space can also produce the valorization of those who experience it through their lived universe, which then reaches the level of place (FERNANDES, 2003; 2006; MELLO, 1990; 1991; 2000; TUAN, 1982; 1998;

2013).

Ilha de Guaratiba, a peripheral part of the western zone of the city of Rio de Janeiro, is currently undergoing a veritable metamorphosis, as its former rural-agricultural configuration is being replaced by today's urban-residential dynamic (FERNANDES, 2006). The growing real estate speculation in vogue in the area began in the 1970s and intensified in the last years of the 1990s. This march of urbanization has transformed Ilha de Guaratiba into one of the veins of Rio's sprawl in recent years. The event in question has stimulated considerable mobility towards the place. The bucolic location, visited sporadically by owners of second homes - a traditional agricultural producer - is undergoing a constant process of land/real estate appreciation and a considerable increase in its resident population. The town in question has for years been presented as the most likely target for volatile speculative real estate capital. Many experts point out that the city of Rio de Janeiro will grow towards Ilha de Guaratiba, especially the wealthier segments (LESSA, 2001; REDONDO, 2012; JANOT, 2013).

In fact, the analysis of spaces and places, through qualitative research, encompasses dimensions concerning meanings, existential questions and even the symbolic and sentimental metamorphoses of those who experience them, recording their way of being and feeling about their world. At this point, it is worth asking: has all this turmoil of changes from rural to urban interfered with the feelings of Guaratibanos in relation to their place?

With regard to the transformations that take place in a given place, we believe that the existential metamorphoses, perspectives and other experiences lived by its inhabitants should not be overlooked. Based on these premises, the past and contemporary geographical contexts of Ilha de Guaratiba will be interpreted taking into account the experiences reported by its inhabitants.

The expansion of Rio's urban fabric, notable for the phenomenon described by Abreu (2008) as "real estate fever," after having covered the coast from the city center (Central Area - South Zone - Barra da Tijuca - Recreio dos Bandeirantes) and, even inland, in the North Zone, "Central do Brasil", Leopoldina and the traditional West Zone, This involves a space to be uncovered, perhaps elevated to the category of home or place, in view of the arrival of new residents who, according to Haesbaert (2004), can either influence the more entrenched groups or be subordinated by them.

With places undergoing intense transformations, it is essential to know from the individuals themselves what their dreams, urgencies, struggles and triumphs are. In fact, the book opts for a shortcut of significant relevance, giving voice to its inhabitants in order to understand their lived universes.

The transition from space to place, as we have discussed, emerges from the experiences lived by individuals and/or social groups in their places. Because it is existential in nature, this subjective metamorphosis occurs from the change in people's attitudes towards their lived world. Because it is part of each individual's intimate collection, this type of experience can be described through testimonies in which the person recounts their daily practices and the material, historical, cultural, affective, social, aesthetic, economic and symbolic values attributed to a once undifferentiated space which - after gaining visibility and/or permanence - is transformed into a place (FERNANDES, 2003; 2006; MELLO, 1990; 1991; 2000; TUAN; 1998; 2013).

By focusing on the transformation of undifferentiated space into place through both the experiences lived in it and the process of attributing value to it, Tuan (2013) refers to the symbolic and affective value (valuation) attributed by individuals and social groups to a given locality. However, the transition from space to place does not only involve the symbolic and affective value of the spatial portion in which the individual is inserted. In many cases, the economic value (appreciation), or that of other spheres, conferred on an artifact, street or area, can represent an indispensable element for qualitative spatial transformations (FERNANDES, 2006).

Places are full of symbols, a precept defended by geographers of the humanistic horizon, according to whom places and their symbols acquire deep meaning according to the emotional bonds woven over the years (MELLO, 2003; TUAN, 2012; 2013). Under these conditions, one of the tasks of the humanistic field is to reconcile, understand and decode the symbolic content of places, since the individual is not distinct from his or her place, as Relph (1976) argues, and each person has a particular and personal geography (LOWENTHAL, 1982; COSGROVE, 2004), a phenomenological approach is needed that privileges the individual in his or her lived world.

The aim of this book is to translate the geographies of Ilha de Guaratiba from the geographical memory of its residents. The approach will start from the testimonies of Guaratibanos and will also seek to uncover their feelings and understanding of the permanence and transformation of their spaces and places. On this path, we intend to decode their personal geographies, forged through their lived experiences, experiences which are responsible for the choice of past and present symbols and for a strong identification with their place.

This research is justified by the need to focus on the individuals and social groups that make up a strategic part of the city of Rio de Janeiro (LESSA, 2001). The area in question, elevated to the status of a place, has been studied by agronomists, biologists, environmentalists and archaeologists in an attempt to highlight its importance in terms of fauna, flora and nature in general. However, Ilha de Guaratiba, like many neighborhoods in the West Zone, lacks research that focuses on the human element in relation to spatial and environmental changes, as well as spaces and places, a field that is specific to the social sciences and geography in particular (MORAES, 2007).

What's more, because it represents my lived universe, Ilha de Guaratiba has always been my research focus in my academic career. I've always lived on Ilha de Guaratiba, being a pioneer and a connoisseur of the area, and I have a strong identification with my lived place. Of course, this fact alone is not enough to justify an academic paper. However, Ilha de Guaratiba is a relevant place to be explored and, under these conditions, to be the focus of studies and research, which is why this work is being developed with greater motivation.

In this way, the research seeks to present a picture of the transformations, experiences, symbolic and affective experiences, among others, that the place in question has been going through, sharpening its study based on the experiences lived by individuals and social groups.

By focusing on the individual(s) in their lived place, this research is based on the phenomenological method, while also using another philosophical support belonging to the niche of philosophies of meaning, namely hermeneutics. The theoretical foundations of this book were sought in the works of humanistic philosophers and geographers who are based on subjectivities, symbolism and identities, therefore, in adherents of the phenomenological method and hermeneutics such as Relph (1976), Schutz (1979); Buttimer (1982), Tuan (1982; 1998; 2012; 2013), Holzer (1992; 2001; 2008), Mello (1990; 1991; 1993; 1999; 2000), among others.

We have tried to maintain this methodological line by basing ourselves on the assumptions of phenomenology and hermeneutics, which seek to decode existential geographies through qualitative analysis where the individual is indispensable and their lived world fundamental (ARANHA, 1996; ABBAGNANO, 2007) in the midst of subject/object indivisibility. Therefore, the proposed methodology is based on bibliographical research and the collection of philosophical material that has provided us with relevant theoretical and conceptual support. Under these conditions, while not violating the principles of humanism in geography, we resorted to interviews and informal conversations with Guaratibanos, where we sought to explain and interpret the geography of the individuals interviewed, in an attempt to decipher the memorable geographies of Ilha de Guaratiba based on and committed to this lived world. In order to achieve a phenomenological analysis, deciphering the feelings and understanding of individuals and social groups in relation to the space they forge, experience and revere, we tried to conduct our qualitative research (interviews) "at the pace of the person being interviewed, that is, freely, spontaneously, informally, without time or subject limitations and in their own environment", as Nogué Y Font (1992, p.90) advocates.

In these terms, we are following in the footsteps of phenomenological philosophy, in which humanistic geography seeks elements to guide its research. The notion most frequently used by humanistic geographers, precisely the concept of place, it is worth repeating, comes from the phenomenological and inseparable idea of the lived world in which there is no separation between subject and object. In these circumstances, the individual and everything around them make up a world full of values, sufferings, dilemmas, joys, experiences, "songs that my mother taught me" (SCHUTZ, 1979, p. 291) and utopias amid friends, relatives, acquaintances, a territorial base, feelings and so on, making up a whole of introjections, strangeness, adherences, reinventions, phobias and belonging (MELLO, 2000).

This book, it is worth repeating, also focuses on translating, decoding and interpreting the geography of Ilha de Guaratiba, giving a voice to its residents, who are the pioneers and deep connoisseurs of their lived world. To this end, the spatial, symbolic and/or existential dimensions of this place are interpreted according to the understanding of its residents. It is also worth pointing out that the many important readings we have done during our academic career have enabled us to take more proficient paths in relation to the method, where new and important conceptual components have been added. In addition, the interviews carried out between September 2014 and October 2015 sparked new elements that were addressed according to these verbal reports. Because they represent a narrative of lived experience, these qualitative studies were essential for unveiling the geography of Ilha de Guaratiba, as well as for decoding the complex web of relationships that involves its residents in their lived universe.

From the process of renewing geography and the emergence of critical geography - understood as the currents that broke with positivist and neo-positivist ideas - geographers of various methodological orientations (existentialists, Marxists, eclectics, etc.) began to take on a more human perspective, seeking through the transformation of the social order, a more generous geography and a fairer space or place, which was organized in function of men (MORAES, 2007). In this vein, we think that the best way to focus on the human being at the center of all things (MELLO, 1991; 2000; 2001; TUAN, 1982; 1998; 2013) would be to use the intimate and private collections of each individual in approaches to their lived universe, where their approach would be the basis of scientific investigation.

With regard to the demarcated space, a geography has been forged, full of feelings and understandings, based on the experiences of the individuals and social groups of Ilha de Guaratiba.

The research is organized into two parts. Part I represents the theoretical and conceptual framework of the book and consists of two chapters. In the first chapter, our aim is to provide the philosophical foundations that will guide this study, as well as the themes and concepts that will serve as the basis for the substantive part of the research (Part II). Therefore, in this chapter, entitled "In search of a more human geography", we look at the presuppositions of the humanistic horizon, establishing this bias as a proposal for the humanization of geography. Complementing our assertion of focusing on individuals at the center of our investigation, we look at some of the principles defended by hermeneutics and phenomenology, which are the main philosophical foundations of humanism in geography.

As we have said, the theoretical framework of this work also extends to the second chapter. As our aim is to establish a multidimensional approach - encompassing the spatial and the symbolic/existential - we wrote this chapter in order to provide a theoretical basis for our empirical reference. "The multiple dimensions of what is experienced" is the title of the second chapter. In this chapter, we seek a theoretical and conceptual basis that encompasses the multidimensionality that permeates our particular lived universes: spatialities, temporalities, artifacts, meanings, changes, permanences, etc.

Part II of this book is about the research itself and consists of the remaining chapters, four in total. In these, we will walk the memorable paths of Ilha de Guaratiba by sharing the geographic memories of its residents. In the third chapter (The investigative scale of an affectively cut-out neighborhood), as the title suggests, we will focus on the characteristics of Ilha de Guaratiba and its residents. In the fourth chapter, our goal is to summarize the process of change in the spatiality in question, seeking to contextualize the geographies of yesteryear in the geographic context of today through the memorial accounts of

Guaratibanos who have experienced and/or are experiencing the transformations that the place has undergone over time. To this end, we used the methodological assumptions mentioned above in order to unveil geographies of yesteryear and contemporary geographies in an approach that privileges the approach of the many informal geographers (LOWENTHAL, 1982; COSGROVE, 2004) of the locality in question, in relation to the current phenomenon in their lived place. That said, in this chapter - entitled "The march of urbanization on Ilha de Guaratiba in the understanding of its residents" - we will describe the aforementioned march of Olympic Rio, with its urbanity flowing into the neighbourhood and the latter into the meanders of Rio's urban gutter, with the aim of gaining a better understanding of the place. The approach starts from the context in which a rural-agricultural panorama prevailed in the locality, passing through the disarticulation of this arrangement of elements and culminating in the transformations that give, today, an urban-residential arrangement to the former peri-urban portion of the city of Rio de Janeiro.

In the fifth chapter, the symbolic content of the place comes to the fore. Through personal accounts, we try to capture and decipher the symbols of the past and present of Ilha de Guaratiba.

Finally, in the sixth chapter entitled The valuation of space promoting the valuation of place, we deal with the experiences and feelings of Guaratibanos in relation to their lived universe which, gaining visibility, has become the center (navel) of their world, that is, with an ethnocentric expression, with regard to geography (TUAN, 1980; 2012). The gradual existential metamorphosis and the other subjective transformations that have taken place in people's lives as a result of the process of change (urbanization) that is taking place in the locality also deserve to be highlighted in this final stage of the book. Therefore, in this chapter, we will try to translate the existential metamorphoses that take place at the heart of the urbanization march that is taking place on Ilha de Guaratiba, as a result, as we have already pointed out, of the Olympic River flowing into the place.

Even though we have provided two chapters containing the philosophical, methodological and theoretical principles of the research, we have tried to observe the articulation between the theoretical-conceptual basis and operational research through the use of themes and concepts that are circumscribed throughout this scientific investigation. It is worth emphasizing that the thematic, philosophical, conceptual and methodological delimitation is implicitly and/or explicitly circumscribed from the first to the last pages of this research. So let's lay the foundations for a memorable geography.

PART ONE: PHILOSOPHICAL BASIS

"Geography is the study of the Earth as the home of people" (YI FU TUAN).

PART I FOR A MEMORABLE GEOGRAPHY

The first part of this book concerns the theoretical and conceptual framework on which it will be based. This intellectual elaboration is the result of a considerable effort made throughout my academic career, having been achieved by reading and analyzing a vast list of authors from various currents of thought during my last fifteen years at university (see references).

Using references from thinkers from different lines of research always represents a risk when it comes to academic research. According to anthropologist Regina Abreu (2011, p. 30-31), however, "it is possible to use authors from different schools and traditions of thought, without this being academic heresy". Professor Abreu stresses, however, that "mixing references to authors in an academic text does not mean indiscriminate use of these authors". Of course, some care must be taken to avoid contradiction. As well as maintaining "a permanent dialogue between theory and research", you shouldn't "lump different authors together, without distinguishing between classic authors and those who have made specific contributions to a given subject" (ABREU, 2011, p. 31-32).

In my research, texts and articles (FERNANDES, 2003; 2005; 2006; 2008; 2009; 2010; 2011; 2012; 2013; 2014 a; 2014 b; 2015), with the exception of those of a purely operational nature (FERNANDES, 2005; 2008; 2014 b) and those related to specific themes (FERNANDES, 2009; 2013), I always sought to dialog with authors of different methodological orientations. The focus of my approaches, however, was maintained, since the philosophies defended by the classic authors that theoretically underpinned my approach were not neglected, nor were they "mixed" with the aforementioned "occasional contributions", as if they had the same weight of importance. This research will be no different, since the approach will start from the philosophical presuppositions of humanism in geography defended in the first chapter by classical and contemporary thinkers of this current of geographical thought (TUAN, 1961; 1980; 1982; 1983; 1986;

1991; 1996; 2005; 2011; 2012; 2013; RELPH, 1970; 1976; 2012; PALMER, 1970; SCHUTZ, 1979; LOWENTHAL, 1982; 1985; 2008; BUTTIMER, 1982; 2015; MELLO, 1990; 1991; 1993; 1999; 2000; 2001; 2002; 2003; 2004; 2005; 2007; 2008; HOLZER, 1992; 2001; 2008; MARANDOLA, 2012).

As this research will also be concerned with uncovering the dynamism of a lived universe in the midst of a veritable march of urbanization, a second theoretical chapter was drawn up in order to demonstrate the multiple dimensions of a lived place (see chapter 2). Among the thinkers who support this secondary chapter are geographers of various theoretical and methodological orientations, including humanists, as well as authors from other fields of knowledge such as philosophers, historians, sociologists, among others. Even though they are not part of the theoretical body of the orientation

The humanistic approach that underpins this text, the contribution of these scholars has been significant in theoretically grounding the place in its multidimensionality.

This first part of the book, made up of chapters one (In search of a more human geography) and two (The multiple dimensions of what is experienced), didn't necessarily need a title. However, imbued with the desire to unveil a place by glimpsing and sharing the geographic memories of its inhabitants, I decided on the title "For a memorable geography". The inspiration obviously came from the term coined by the geographer recognized, along with Tuan, as the father of humanistic geography: David Lowenthal (HOLZER, 1992).

According to Lowenthal (1982), "memorable geographies" are those made up of interpretative texts, where the past plays a fundamental role in the construction of the particular universes of individuals and social groups. In this context, lived and shared experience becomes the foundation of a new way of knowing and doing geography. But how can an individual share their experiences without referring to their memory? As such, memory is a concept (GONDAR E DODEBEI, 2011; SILVA, 2015) or category (ABREU, 1998) that is fundamental for revealing the personal and/or collective geographies of different places.

In this approach, memory is represented by the ability to store and/or save information related to our past: remote or recent (ABREU, 1998). Memories are the memories that refer us to different temporalities and spatialities, these being the two dimensions of both social construction and human experience (SACK, 1986). By referring us to personal and collective identities, memory becomes one of the dimensions of places (SILVA, 2015). A common resource in both historical and geographical studies, qualitative research seeks to sharpen individuals' memories by asking them about events and/or places of yesteryear (FERNANDES, 2015).

This research is no different, as it is based on the memories of the residents of Ilha de Guaratiba, memories linked to different past contexts experienced by Guaratibanos. As well as being based on the historical and geographical memory of the aforementioned residents, this study, it is worth repeating, is also based on the assumptions of humanism in geography - as well as on a multidimensional approach (urbanization march, symbolic and affective experiences, among other facets that involve the dynamic lived universe). Thus, following the definition in my favourite dictionary (BORBA, 2011), a "memorable" geography would be one that is retained in the geographic memory of the residents of the place being researched, a "humanized" and unforgettable geography, par excellence, that deserves to be unveiled (FERNANDES, 2014).

In view of the above, let's consider the pertinent questions regarding the presuppositions of the humanistic horizon, as well as the principles defended by hermeneutics and phenomenology, which are the main philosophical foundations of the current that guides this work.

1 IN SEARCH OF A MORE HUMAN GEOGRAPHY

In Lowenthal's view (1982, p. 137), a memorable geography would not be tied to mere compendium texts - much less to generalizing perspectives that ignore particularities, individualities and singularities - "but to interpretative studies that incorporate a strong personal point of view". Following this premise, we understand that a geography committed to universalizing aspects that encompass society as a whole, excluding the particular universes of individuals and social groups, cannot account for a genuinely human and personal geography, forged by each informal geographer in their lived world (LOWENTHAL, 1982; COSGROVE, 2004).

By underestimating the rich material represented by the experiences of the lived world, as well as the conceptions that derive from these experiences, "geography went a long way towards introducing man as a thinking being into its research" (MELLO, 1991, p. 1). At the beginning of the 1970s, however, some geographers, frustrated with a geography in which the human being represented just another element of the studied landscape (MORAES, 2007), "began to look to philosophies of meaning for answers to their anxieties" and for ways to break away from the positivist and neopositivist assumptions that predominated in geographic science (MELLO, 1990, p. 22). Humanistic geography emerged, a perspective that focuses on man at the center of all things, since every human being thinks and philosophizes, and is therefore capable of reflecting on the phenomena of lived worlds (BUTTIMER, 1982; MELLO, 2000; RELPH, 1976; TUAN, 1982).

A more humanistic geography is one that reflects on geographical phenomena with the aim of achieving a better understanding of man and his condition (TUAN, 1982). In this sense, the humanistic approach to geography, with existential phenomenology as its underlying philosophy and hermeneutics as its method of interpretation (BUTTIMER, 1982; GOMES, 2007), seeks to value the experience of the individual or social group, aiming to understand the mosaic of feelings and people's understanding of their places. From this perspective, the geographers of the current in question argue that their approach deserves the label of humanistic, since they study aspects of man such as meanings, values, goals and purposes (CHRISTOFOLETTI, 1982), as well as allegories, dreams, daydreams and reminiscences (MELLO, 1991; FERNANDES, 2014).

1.1 Another horizon in search of the humanization of geography

The humanistic movement highlights man and treats him with his meanings, values, objectives, dilemmas and actions, as opposed to the abstract, mechanistic and deterministic approach of previous paradigms. Criticism of the reductionist view of man, especially after 1970, helped humanistic geographers to interpret feelings and understand the relationships between men and their lived world. By advocating the incorporation of the subjective dimension and the lived experience of individuals and social groups into their approaches, this perspective proposes an understanding of the human world through the study of people's relationships with nature, their geographical behavior, as well as their feelings about space and place (TUAN, 1982; 2013).

The ideas mentioned above are relevant to this geographical trend. In the place, the individual finds himself at home and even integrated. This conceptual expression makes up this world full of feelings and affections, a center of significance or a focus of man's emotional action. Place is not just any location, but one that expresses affectivity and values for the individual or their community (CHRISTOFOLETTI, 1982). In contrast, space is represented by any portion of the earth's surface, being vast, unknown, undifferentiated, rejected or even hated (MELLO, 1990; 1991; 2001; TUAN, 1982; 1998; 2013), possibly to be captured or conquered.

Despite the distinction between these key concepts in geography in general (CORRÊA, 2002) and the humanistic branch in particular (HOLZER, 2001; 2008), in experience, the meaning of space can merge and/or be confused with that of place. What begins as an undifferentiated space becomes a place as we get to know it better and give it value (TUAN, 2013). According to Mello (1990, p. 105):

> Certain spaces only become places after a long experience. What is initially ugly or even hated, over time takes on the status of a place. Spaces become Places because of contact with other people and because of emotional, economic and other exchanges.

Humanistic geography does not neglect the affective dimension in the midst of the lived/existential. As such, it is fundamental for humanistically inspired geographers not only to approach the "spatial distribution of social facts, but also the way people live in the places where they live or visit, drawing an experience from them" (CLAVAL, 2001, p. 46). Along these lines, "being together, being close, does not mean physical proximity, but an affective relationship with another person or place" (CHRISTOFOLETTI, 1982, p. 23).

Places and people that are physically distant can be affectively very close. Therefore, the study of spaces and places goes back to the analysis of the feelings and spatial ideas of people and groups of people (TUAN, 1982). "From a positivist perspective, geography only concerns the analysis of spatial organization. Under the humanistic horizon, space and place take on very different characteristics" (CHRISTOFOLETTI, 1982, p. 23), and it is up to the humanistic geographer to translate what they represent into a coherent structure (CHRISTOFOLETTI, 1982; TUAN, 1982).

Valuing attitudes leads to a concern with tastes, preferences, characteristics and particularities of places. It also values "the environmental context and the aspects that result in the charm and magic of lived universes, in their personality and distinction". It is then the intertwining between the person or social group and the place (CHRISTOFOLETTI, 1982, p. 23), since the individual is not distinct from their lived world (RELPH, 1976).

Tuan (1982, p. 159), for his part, reports that "the contribution of humanistic geography to science lies in revealing materials of which the scientist, confined in his own conceptual bubble, may not be aware". For the researcher:

> This material includes the nature and range of human experiences and thoughts, the quality and intensity of an emotion, the ambivalence and ambiguity of values and attitudes, the nature and power of symbol and the characteristics of human events, intentions and aspirations.

In order to elucidate this mosaic of material and immaterial objects in the midst of subject/object indivisibility, the humanistic geographer..:

> You should have a penetrating interest in philosophy, because it raises fundamental questions of epistemology for which we can seek explanations in the real world. Philosophy also provides a unified point of view from which a whole range of human phenomena can be systematically evaluated (TUAN, 1982, p. 161).

In these terms, we are talking about hermeneutics, which according to Mircea Eliade (1971), is the only effective method of interpretation that humanism cannot do without, and phenomenology, in which humanistic geography seeks elements to guide its research (GOMES, 2007).

1.2 **Philosophies of meaning: methodological foundations of humanistic geography**

Based on the principles of phenomenology and hermeneutics, humanistic geography is interested in understanding the soul of places based on the lived experiences of individuals and social groups. This perspective believes that place is an integral part of being, and that each individual is an informal geographer qualified to discuss the soul of places, since it is man who produces, learns, lives and transmits geography (BUTTIMER, 1982; COSGROVE, 2004; LOWENTHAL, 1982; MELLO, 2000; SCHUTZ, 1979).

"Phenomenology is the philosophy present in the largest number of humanistic studies in geography" (MELLO, 1991, p. 36), and is also considered a research method (ARANHA, 1996; GOMES, 2007). "Its creator, the German philosopher Edmund Husserl (1859 - 1938) criticized scientific theories, particularly those of positivist inspiration, which were excessively attached to objectivity and the belief that reality is reduced to what is perceived by the senses" (MELLO, 1991, p. 36).

Etymologically, phenomenology is the study of phenomena, and its role is to analyze the dynamics that provide meaning and significance to objects, treating the world and human beings in an inseparable way (SCHUTZ, 1979). By examining the concept of phenomenon, which in Greek means what appears (ARANHA, 1996), we can better understand that phenomenology deals with knowledge as it appears, that is, as it presents itself to consciousness. As a source of Meaning for the world, consciousness is not restricted to mere intellectual knowledge, but is the generator of intentionalities that are not only cognitive, but also affective and practical. Looking at the world is the act through which man experiences it, imagining, judging, loving and fearing it (ARANHA, 1996; SCHUTZ, 1979).

Phenomenology criticizes the naturalistic tendency that guides the method of the human sciences. For this philosophy, there are no facts with the objectivity sought by positivism, since we don't understand the world as a raw fact, devoid of meaning. The world I decode is a universe for me, hence the importance of feeling, understanding and the network of meanings that surround our lived universes (ARANHA, 1996). For Schutz (1979), the irreducible starting point for phenomenological foundations concerns the experiences of conscious human beings, who live and act in a world that they capture, interact with and interpret - assuming multiple meanings.

Everything experienced by human beings corresponds to experiences in/of their lived world. They constitute it, they are directed towards it, they are tested and experienced in it. The lived universe is simply the whole sphere of everyday experiences, directions and actions through which individuals deal with their interests. According to phenomenological philosophy, each individual constructs their own world. Thus, subjectively, no two people could experience the same situation in the same way (SCHUTZ, 1979). In this respect, Buttimer (1982, p. 167) points out that:

> Phenomenologists have been the most systematic spokespeople for a combined effort to reconcile heart and mind, knowledge and action, in our everyday worlds. Challenging many of the premises and procedures of positive science, they set out a radical critique of reductionism, rationality and the separation of subjects and objects in empirical research. With the existentialists, they argued for the liberation of lived experience, calling for more concrete descriptions of space and time, and their meanings in everyday human life.

Since its common feature is the inseparability of subject and object, phenomenology radically examines the phenomena of consciousness or lived experience, looking for the facts as they are produced, interpreting the apprehension of the essence and in this way investigating the acts and understanding of the lived world. In this context:

> Place emerges as a key concept in humanistic geography stemming from the phenomenological notion of an emotionally lived world, shaped, introjected and coated with events, people, itineraries, struggles, ambiguities, involvements, dreams, follies, "songs my mother taught me", a territorial base and all sorts of elements that allow a person to feel at home or, on the other hand, distanced in the midst of a topophobic estrangement (MELLO, 2005, p. 34).

> For a long time, geographers excluded neighborhood ties, knowledge, pleasantness, topophobia, attachment to spaces and places, everyday experiences and the links between people and the environment from their approaches. Taking these attributes into account, phenomenology serves as a bridge to these specialists with a view to understanding the lived world, because - unlike science which omits life issues - it does not treat the world independently of human beings (...). On the basis of the lived world, the geographer can understand how the magic of places arises, the intrinsic particularities

of each territorial portion, the distinction between different parts of the city, enchantment, contempt, attraction and what is typical of places (...). The lived world is the consciousness and intimate environment of each person, emotionally shaped and coated with events, relationships, ambiguities, involvements, values and meanings, which comprises human beings with all human actions and interests, labors and sufferings (MELLO, 1991, p. 37-38).

Each person's lived world existed before they were born, and they experience and interpret their world based on values and a stock of personal experiences, as well as through other individuals who pass on past and current knowledge (MELLO, 2000, p. 57).

Intersubjectivity, or the interworld, is the world shared by different people, the scene and object of human actions and interactions. The lived world, continuously experienced, is modified by human actions, which also modify their actions (MELLO, 1991). "The stock of experiences is a daily enrichment, both practical and theoretical, which provides man with the elements to act and think" (MELLO, 1991, p. 38). "However, this knowledge is not homogeneous but incoherent, partial, contradictory and ambiguous" (MELLO, 2000, p. 58). "Knowledge of the world, received through formal and informal culture and supplemented by personal experience, generates intimacy and affection for the place lived" (MELLO, 2000, p. 58). This world, which for phenomenology is the context within which consciousness is revealed, occurs to our experience and interpretation (BUTTIMER, 1982; ELIADE, 1971; GOMES, 2007; MELLO, 2005).

"Hermeneutics, another philosophy of meaning, used by humanistic geographers, has as its precursor the German Wilhem Dilthey (1833 - 1911) who added to this philosophical movement - close to Schutz's phenomenology - elements of interpretative importance" (MELLO, 1991, p. 41). The origin of the term lies in antiquity, inspired by the Greek mythology of Hermes, god of communication, in charge of bringing messages from Olympus (GOMES, 2007). Originally used by ancient theologians as a methodology for interpreting the Bible, the term later came to designate any effort to scientifically interpret a difficult text that requires an explanation. Nowadays, hermeneutics is a philosophical reflection on the interpretation or understanding of symbols, and is fundamental to all humanities and all disciplines concerned with the interpretation of human works (ABBAGNANO, 2007; JAPIASSÙ E

MARCONDES, 2006; PALMER, 1970).

"In hermeneutics", as in phenomenology, "there is no separation between subject and object". As such, this "interpretative philosophy attempts to explain the contents of the mind and other aspects of lived experience" in the midst of introjections, where individuals and social groups are not dissociated from their experienced territorial base. It is up to the geographer to clarify the meaning of concepts, symbols and aspirations, since all of this is linked to space and place. In other words, "the humanistic geographer's task is to interpret the ambivalence, ambiguity and complexity of individuals' and/or social groups' awareness of the environment" (MELLO, 1991, p. 42).

It is worth noting that "the boundaries between phenomenology and hermeneutics are not very rigid". Thus, "several humanistic geographers - including Tuan, Buttimer, Lowenthal and Relph - although they classify themselves as phenomenologists, exhibit the hermeneutic movement in an unmistakable way" (MELLO, 1991, p. 42). Following this line of thought, this study endeavors to try and translate the personal (individual and collective) geographies of Ilha de Guaratiba, using the philosophical assumptions presented here. First, however, it is worth pointing out that - like any place - Ilha de Guaratiba has multiple dimensions. In order to uncover the dynamism in which it is enveloped, its multiplicity needs to be considered. But which dimensions are we talking about? In search of answers, let's consider the following elucidations.

2 THE MULTIPLE DIMENSIONS OF LIVING

As we noted in the previous section, the humanistic orientation emerged as a reaction to positivist and neopositivist assumptions, where the separation between subject and object is clear. Furthermore, in positive science, quantitative methods are prioritized, and the followers of these approaches are only interested in materiality, leaving subjectivity, symbolism and existential issues linked to the experiences lived by individuals and social groups absent from their studies (MELLO, 1990; FERNANDES, 2014).

With the advent of interpretive perspectives, the different meanings inherent in the lived universe of social groups became the focus of geographical research (CORRÊA AND ROSENDAHL, 2012). In this context, geographers began to focus on the bonds that individuals weave between themselves and their lived universe, as well as why people attribute meaning and subjective values to their places (CLAVAL, 2007).

In the words of Gomes (2007), one of the main characteristics of humanism in geography is its holistic nature, looking at the whole in order not to lose the richness of the whole. According to Holzer (2008), the idea of a discipline centered on the study of human action and imagination and the objective and subjective analysis of its products, which aimed to constitute a science of synthesis that went beyond Cartesian and positivist parameters, takes us back to the 1920s. Following this line of thought, in 1947, Wrigth - through the concept of geosophy - established the basis of a science project that embraced the different approaches, as well as the objective and the subjective indistinctly (WRIGTH, 1947; LOWENTHAL, 1982; HOLZER, 1992; CASSIRER, 2011).

To paraphrase Almeida (2010), every place has an objective and a subjective dimension. Both together make up what is experienced. The immaterial dimension, however, is the one that provides the connotative component that also ends up being an inseparable part of the concrete dimension (ALMEIDA, 2010).

In the course of the process of renewing geography, it is common to see both a break with previous patterns (GOMES, 2007) and attempts to reconcile polarized positions (ALMEIDA, 2010). As Claval (2007) points out, materialist geographers are interested in spatial structures and humanist geographers in individual and collective initiatives and actions. For Claval, the ever-increasing integration between the subjective and objective dimensions of place is an example of analysis on the middle road between two approaches: the spatial and the existential; urbanism and humanism; the material and the immaterial, etc.

In turn, Antonello (2010), seeking a synthesis between the objective and subjective dimensions, points out that place is where lived experiences are anchored. The author also states that geography must seek to overcome the limits imposed by the division of scientific work, seeking to follow a path that leads to a plurality of focuses of analysis, in order to understand the complex reality that surrounds us (ANTONELLO, 2010).

Indeed, Haesbaert (2005), another geographer who defends the inseparability between the objective and subjective aspects, proclaims in his research that territoriality arises as a result of the intertwining of the different dimensions of reality. The author warns that it is precisely by making a rigid separation between territory as domination (material) and territory as appropriation (symbolic) that many ignore the complexity and richness of the multi-territoriality in which we are involved (HAESBAERT, 2005).

According to Bastos (1998), the real is not only made up of the material universe, but also of the meanings derived from the relationship between individuals and their lived world. For the researcher, in the apprehension of geographical space there is a concrete dimension (the production of material space) and a symbolic dimension (representations). These two dimensions of reality interact with each other in a symbiotic (subject-object) and inseparable relationship. In this sense, spaces and places can be represented according to an imaginary in which their materiality must not be denied. Along these lines, there is no justification for separating the researcher's subjectivity when interpreting the place under study, since it presents itself in such a way that the concrete and the abstract are inseparable parts of the same reality. In the dynamic lived universe where man is the geographic actor and the place his niche of belonging and collective memories, there is an imbrication between the material and the symbolic, between the spatial and the existential, between the objective and the subjective (BASTOS, 1998).

Like the aforementioned researchers, Cosgrove (2003) advocates a geography in which the objective and subjective dimensions of reality are taken into account. For this geographer, human beings promote spatial transformations with their sensory and material reality. In this sense, all human activity is material and symbolic at the same time. Despite being symbolically constituted, the lived world is material and should not deny its objectivity (COSGROVE, 2003). The lived experiences and other subjective experiences of individuals and social groups occur in a context and should not be seen as independent of materiality (CORRÊA, 2003).

According to Geertz (2013), in anthropology - likewise - there is an endless debate about whether culture is subjective or objective. For the author, culture should be understood as an interwoven system of interpretable signs (symbols), as a context where representations and substantive content are intertwined. In relation to this unnecessary complication, related to the dualism between the concrete and the symbolic, Geertz points out that he has tried to resist excessive subjectivism and tried to keep the analysis of symbolic forms considering both dimensions of reality (GEERTZ, 2013).

In line with the researchers cited in this section (CORRÊA, 2003; COSGROVE, 2003; BASTOS, 1998; HAESBAERT, 2005; ANTONELLO, 2010; ALMEIDA, 2010; CLAVAL, 2007; WRIGTH, 1947;

LOWENTHAL, 1982; GEERTZ, 2013) I believe that a multidimensional approach is possible in geographical studies, where - equally - the different facets of the lived universe are considered. Wasn't it in the gap left by exacerbated objectivism and the consequent neglect of the subjective values of individuals and social groups, linked to positivist and neopositivist paradigms, that humanism in geography emerged?

In my opinion, the result of the divorce between the objective and subjective dimensions would be the construction of a geography that does not share the aspirations of a science that was born with the ambition of unveiling the world in its totality and diversity (SILVA, 1988; CLAVAL, 2007; MARANDOLA JR, 2012).

In the following topics, we will look at some concepts and themes that can contribute to approaches that focus on place in its multiple dimensions, starting with one of the main themes of this research.

2.1 **On the memories of places**

As we have argued in this research, places have many dimensions of meaning (BUTTIMER, 2015). One of the ways pointed out by different authors in order to explore the dynamic and multifaceted lived universe would be through a glimpse into the intimate and private collections of those who experience it. However, as every lived experience inexorably goes back to the past, if we want to follow this path, we will need to rely on the individual and collective memories of individuals and social groups (LOWENTHAL, 1982; 1985; 1998; ABREU, 1998; MELLO, 2002; HALBWACHS, 2013; BUTTIMER, 2015).

No matter how much a place changes, for example as a result of urbanization, it will always retain a residual content of past temporalities. There is, however, a range of events that leave no physical traces. These are often recorded in the memory of the individuals who follow these changes. What we think and feel about our geographies is obviously based on experiences from somewhere in the past. From this link, everything that happens on our experienced ground becomes part of our very existence (LOWENTHAL, 1998). Everyday life is based on place, which is the result of all our moments, the sum of our memories and the product and combination of all our lived experiences (MENDILOW, 1960).

Possessed by the past, Lowenthal admits to living in a tangle of eras. For Lowenthal, the past resides in us through our memories and recollections of different moments (LOWENTHAL, 1985; 1998). By populating the thoughts of human beings, the past is alive in our memory (HIGHET, 1949). In fact, the scenarios and experiences glimpsed and lived become part of our memory (BUTTERFIELD, 1965).

"All awareness of the past is based on memory. Through memories we regain awareness of previous events, distinguish between yesterday and today, and confirm that we have already lived through a past" (LOWENTHAL, 1998, p.75). Memory is both personal and collective. The memories of individuals and social groups sustain their sense of identity in relation to their experienced ground. Our lives are impregnated by our memories. At every moment, we bring back some event from the past (LOWENTHAL, 1998).

Memories tend to accumulate with age. Our stock of memories increases as life goes on and experiences multiply. Although there are individual and collective memories, memory is always personal. Memories cease to be personal only when we decide to share them (LOWENTHAL, 1998). The memory of the place is inscribed not only in some artifacts from the past that persist in its landscape as geographical testimonies, but - above all - in the memories of the experiences lived by men and women in their lived universe. In this context, "place transcends materiality, although it is not dissociated from it, because spatial memories are eternalized in our memories" (MELLO, 2002, p. 63).

2.2 Spatialities and temporalities

When faced with a piece of data or information about an event, two questions are inevitable: when did it happen? In what place? These questions are due to the fact that every event affects specific spatialities and specific temporalities (SANTOS, 2002). Thinking of place as a composite of spatialities and temporalities means recognizing its dynamics, generating and involving forms, functions, contents (ABREU, 2003) and meanings (TUAN, 2013) inserted in multiple levels of investigation.

For a long time, geography favored spatial analysis without paying due attention to time (SALGUEIRO, 2003). Nowadays, however, the spatial and temporal dimensions have to be valued (ABREU, 2003). The categories of spatiality and temporality obviously lead us to questions about space and time (GEIGER, 2003). However, spatiality goes beyond the rigidity associated with space as a mere stage for events, representing an advantage when approaching the spatialization of a dynamic phenomenon such as urbanization, for example. Similarly, we can talk about temporality by referring to different periods (SALGUEIRO, 2003; FERNANDES, 2010). In geographic approaches, it is not enough to unveil the multiple dimensions of contemporary spatiality. It is also necessary to investigate the past temporalities responsible for the legacies already configured. This is the path that geographers must follow if they want a geography that is genuinely committed to spatiality (ABREU, 1997).

Space and time are basic categories of human existence (HARVEY, 1992), since they combine different temporalities. In these approaches, it is essential to introduce temporality into space. In this reading, space and time are not reducible to each other. They are distinct but complementary terms (MASSEY, 2008).

The actions of nature and human beings are inscribed in space (spatiality) and time (temporality). The phenomena derived from natural, social and existential influences are spread out, located in spaces and places through their limits and dimensions. Through these actions in space and time, different places emerge. Spatialities, as well as temporalities, emerge from the different actions of nature and human beings on the earth's surface (CORRÊA, 2011).

In the words of Cosgrove (2004, p.110), "above all, a historical and contextual sensibility on the part of the geographer is essential. We must resist the temptation to dislocate the landscape from its context of time and space".

The spatial and temporal dimensions of lived experience have always been at the heart of the humanistic perspective in geography. Both Lowenthal (1982; 1985; 1998) and Tuan (1980; 1983; 2011; 2013) - researchers identified by Holzer (1992, p. 489) as "the undisputed fathers of humanistic geography" - favored these categories in the studies that underpinned this current of geographic thought. According to Lowenthal (1982), time exerts a strong influence on our individual perspectives in relation to our particular universes. For Lowenthal, "each particular world has had a career in time, a history of its own" (p.138), and in this sense, "every personal history is the result of a particular environment" (p.139). The relationship we have with our lived universe is built on our lived experience, and this takes time. Lowenthal concludes his mental elaboration by pointing out that "all kinds of experiences (...) come together to make up our individual picture of reality" (p.141). In the course of time, "as artists", we can create and/or organize spaces and places (LOWENTHAL, 1982, p.141).

Tuan (1980; 1983; 2011; 2013) also gives undeniable importance to the spatial and temporal dimensions, elevating them to the level of the "framework" of humanism in geography (TUAN, 2011, p.8). In his conception, Tuan points out that "in the imagination it is easy to treat space" and "time separately". "In lived experience", however, "they are inextricably linked" (TUAN, 2011, p.8). "Movement requires time and takes place in space" (p. 15). "The longer we stay in a place, the better we know it and the more deeply meaningful it becomes to us" (p. 17). For Tuan (2011, p. 18), space and time "are overlapping categories of human experience. If they are not considered together, the world of geographers will retain an air of unreality, abstracted from life as it is lived".

Time is implicit in all spaces and places. We all have a sense of space and time. "The ease with which we confuse spatial and temporal categories is evident in language." "Length is often given in units of time" and "the passage of time" is described as length. "Time is still volume": the great moments of life (TUAN, 2013, p.147). "Space is history". In this sense, "space and time have always been structured according to individual human feelings and needs" (TUAN, 2013, p. 152-153). "Space has temporal significance", both "in the poet's reflections" and "at the level of everyday personal experiences" (TUAN, 2013, p. 156). "Space and time coexist, intermingle and each is defined according to personal experience" (TUAN, 2013, p.161). "We owe everything we are to the past. The present also has value, it is our experiential reality" (TUAN, 2013, p.239).

In his reflections in search of a geography that is not limited to scientific and academic rigors, Wright (1947) indicates the need for geographers to be concerned with the study of both spatiality and geographicity. The latter is understood as the deep involvement of individuals and social groups with their lived universe, which is the intimate spatiality of each human being. In this sense, both spatiality (the characteristics and dynamics of places) and geography (experiences, meanings and symbolisms) are linked to our existence. For this reason, it is worth repeating that the ambivalence and complementarity of these essential elements of reality should not be overlooked (WRIGHT, 1947; DARDEL, 2011).

2.3 **Polyvocality: the multiple interpretations of the lived universe**

There is no ignoring the subtle and complex ties that bind human beings to their places. The symbolic and other subjective connotations are present in the practice of interpreting and analyzing spaces and places in their meanings. The human being is an informal geographer (LOWENTHAL, 1982) and the place is his or her lived world in which relationships are mixed up in a tangle of ties, experiences and clashes, where personal feelings, collective memories and symbols are present. From this point of view, we can understand place as a reservoir of symbols to be interpreted, both by ordinary individuals and, consequently, by geographers. The place, as a humanized shelter, thus allows for multiple readings concerning different manifestations that express as many interpretations as there are subjective and intersubjective meanings.

Human places are made up of several layers of meaning. These multiple layers turn the lived universe into a symbolic place open to interpretation. From this perspective, "geography is everywhere" (COSGROVE, 2004, p. 96) and each individual, by thinking about and seeking to understand their lived world, becomes an informal geographer (COSGROVE, 2004; LOWENTHAL, 1982), able to interpret their particular universe.

According to Lowenthal (1982, p. 105) "anyone who examines the world around them is, in some way, a geographer". For Lowenthal, specialists in this field need to consider that there are other people decoding the world from different perspectives and that the view of the lived universe should not be obtained from just one perspective. Each conception of the world is unique, and there is no possibility of (LOWENTHAL, 1982; CORRÊA, 2007). At this rate, one of the reasons why such differentiations about what is experienced are unique is because all information is inspired and edited by the individual's sense of place (LOWENTHAL, 1982). These are an immensely rich accumulation or sum of information about individuals and social groups.

According to Geertz (2013), man is tied up in webs of meaning that he weaves himself. In order to decode them, scientific laws must give way to interpretative philosophies that can translate the range of expressions inherent in all human activity, which is both material and symbolic (COSGROVE, 2003; GEERTZ, 2013).

Based on Hall (1997), Corrêa (2007, p.5) states that "symbols are open to different interpretations, each based on the experience, values, beliefs, myths and utopias of the social group that interprets them". The meanings attached to symbols vary from person to person. This condition promotes polyvocality, i.e. different interpretations of the same symbol. Polyvocality becomes a kind of antidote to an imposed, single or unilateral meaning (CORRÊA, 2007).

The revelation of the occurrence of multiple layers of meaning linked to human relationships (GEERTZ, 2013), as well as their overlapping, has long been neglected (PANOFSKY, 2004), pointing to a pressing need: approaches that privilege the human element in its multiplicity, since place is multidimensional.

2.4 The place in its multidimensionality

According to Relph (2012), place has been a concern of philosophers since classical antiquity. Plato saw it as "the nourishment of being, while others approached it from a geographical sense as the context in which beings are gathered together" (RELPH, 2012, p. 18). From the 17th century onwards, however, the Cartesian conception of space as a measurable dimension excluded the place of philosophy and the sciences. Only in the 20th century did phenomenological philosophers - "especially Husserl, Heidegger, Merleau-Ponty" (RELPH, 2012, p. 18) - identify the profound inadequacies of Cartesian logic. In the view of these thinkers, positivism promotes the incision of philosophy by leaving out the feelings, emotions and experiences of human beings. By reducing spaces and places to a single (material) dimension, positive science promoted a deficient geography (RELPH, 2012).

The aforementioned indifference to the subjective dimension of human existence is, according to Relph (2012), one of the possible reasons for the recent interest in place. However, the aforementioned geographer does not fail to point out that the concept in question has multiple dimensions. For Relph, as well as bringing together the qualities, experiences and meanings of our experience, place has other subjective aspects (spirit, meaning, rootedness, interiority...) as well as objective ones (location, physiognomy, buildings...). The essence of place presupposes the observance of its multidimensionality (RELPH, 2012).

In her debate on the significant dimensions of place, Livia de Oliveira (2012, p.12) points out that "every place acquires identity through its various spatial dimensions: location, direction, orientation, relationship, territory, spatiality and others". In the text in question, Professor Oliveira seeks a junction between the physical, experiential and symbolic dimensions in order to sustain the different spheres or layers of meanings "that cohere to the constitution of place" (MARANDOLA JR, 2012, p. 15).

In their laborious reflections on place and the subject, Berdoulay and Entrikin (2012) agree that the intense feeling of belonging, which creates a fusion between the individual and their lived world, can be promoted by measurable elements or by ephemeral existential relationships that make up our daily lives. For the aforementioned thinkers, by promoting changes in their place, human beings transform the Earth into their world. These changes, in turn, also affect individuals and social groups in their relationships. In this sense, the subject (human being) and the place (objective and subjective reality) form an inseparable whole (BERDOULAY AND ENTRIKIN, 2012).

Place - the lived universe par excellence - goes beyond geographical objectivity, which ignores the subtleties responsible for its evaluative aspects (GALLAIS, 2002; FERNANDES, 2010). The lived experience also carries with it a dimension richer in subjective and existential aspects. It is a sphere of recognition and familiarity pertinent to everyday life, where its significance deepens its (geo)symbolic role (BONNEMAISON, 2002).

In the words of Dardel (2011), geography, the work of man, displays a constructed space that has meaning. These constructions exalt man by interweaving his customs, habits, behavior, ideas and feelings into his environment and horizon. The different constructions differ in quality and meaning. Among the most important elements of built space is man's habitat, his home, loaded with existential and symbolic values (SCHUTZ, 1979; TUAN, 2013; DARDEL, 2011). Place is the synthesis between man and the Earth in its subjectivity, values and meanings. This geography represents, according to Dardel (2011), "the nature of geographical reality".

Despite its multiplicity, place is unique. As subjects, human beings construct places of belonging and identity and are, at the same time, shaped by these places. This exchange between subject and object, between individuals and their places (inseparable elements), "builds obstacles to the postmodern and metropolitan tendency to see each place as the summary of all the others" (BERDOULAY AND ENTRINKIN, 2012, p.112).

Place has several dimensions (DARDEL, 2011), representing the basis of human existence and the center of meaning, a postulation suggested by Entrikin (1980). In Bachelard's (2008) reflections, place carries the essence of the notion of home and the topography of our intimate being. In another school of thought, Santos (2002) formulates place as follows: the obligatory depository of the event and the irreplaceable theater of human passions. In this context, it becomes the basis for the reproduction of life and a product of human relations, according to Carlos (1996). At the same time, place, in the words of Armando Corrêa da Silva, gains the property of a spatial dimension in its entirety and totality of what could be assumed to be reality (SILVA, 1988). According to the canons of the humanistic perspective, space endowed with value and meaning emerges at the level of place or home (TUAN, 2013). By reading its multiple definitions, place can be considered a lived universe, not only in its concreteness and objectivity, but with all the partialities of human imagination (BACHELARD, 2008).

In its multidimensionality, place is a hybrid entity made up of materiality, symbolism, spatialities, temporalities and the existential way of experiencing it. Reconciling and decoding this multiplicity represents the nature of geography (DARDEL, 2011), the task of which I set myself the task of developing throughout this thesis.

Individuals and social groups, it is worth repeating, are not distinct from their lived places, in the midst of the phenomenological inseparability between subject and object, since each person has an individual geography, thus being a kind of informal geographer. In this regard, our goal is to decode the lived experiences of Guaratibanos, woven over time into their common territorial base, in an attempt to address the spatial / existential transformations in Ilha de Guaratiba (BUTTIMER, 1982; LOWENTHAL, 1982; COSGROVE, 2004; FERNANDES; 2010; 2014 a).

Through our informal interviews, the residents of Ilha de Guaratiba expressed their views on the changes that have taken place in their communities, as well as the existential metamorphoses that this phenomenon has caused. These accounts are the foundations we will use to approach the process of spatial transformations in Ilha de Guaratiba. By reading and interpreting them, we will try to elucidate the complex spatial and existential dynamics of the locality, since the concept of place undertaken here brings together - inextricably - form-context, object-subject, presence-existence, past-present, the Guaratibanos and their experienced ground.

Starting from this brief thematic and/or conceptual exposition, we intend to use phenomenology and hermeneutics to unveil the geographies of yesteryear and the contemporary geographies of this place. The memories, in short, the experiences lived and shared by the residents of Ilha de Guaratiba - exposed through verbal reports - will guide us in the following approach.

PART TWO: OPERATIONAL RESEARCH

The midwife who helped me to be born must have buried my navel in this piece of ground. Maybe that explains why I'm so attached to this place (Evanir de Souza - resident of Ilha de Guaratiba).

PART II AN ISLAND OF MEMORABLE GEOGRAPHIES

As a common trait, individuals are excellent connoisseurs of the dynamism of their lived universe (BUTTIMER, 1982; 2015), and are therefore able to reveal the nuances of their affective and frequency domains more aptly. In this respect, they draw on their geographic memories, since their lived experiences undoubtedly refer to their past experiences. In order to share something related to their place or bring back experiences they once had there, women and men resort to their memories, mixing them with the present.

With regard to the relationship between the geographical concept of place and the memories of its inhabitants, let's consider the opinion of the humanistic geographer Joâo Baptista Ferreira de Mello, according to whom

> the concept of place, traditionally explored by geographic knowledge, is intertwined with the trajectory of humanistic contribution, assuming, within this horizon, the meaning of home, as it is full of experiences and resourcefulness, at the same time, a respite of stability and well-being, as well as a welcoming home and an introjected field of movement and belonging to be defended. These principles were learned and adapted from phenomenological philosophy and from the dynamism of the lived world, an inseparable whole made up of people, land, belongings, events, acquaintances and all sorts of elements that allow the individual to feel at home (...). In this whirlwind of permanence, understanding, exchanges, actions, conflicts, dreams, delusions and affectivity, our geographies are perpetuated, because place transcends materiality, although it is not dissociated from it, because spatial memories are eternalized in our memories (...). Human geography is concerned with the organization of space. The geographers of the humanistic wing do not deny this perspective, reworking the concept of place based on the lived experiences, feelings and understanding of individuals and social groups, pointing out its multidimensionality and the various ways of understanding it (MELLO, 2002, p. 63).

As Mello points out in the quote above, the phenomenological notion of place that underpins humanistic thinking considers the individual and their territorial base to be inseparable. Places, as we know, are the result of distinct temporalities and spatialities. In order to achieve a better understanding of the place or to restore places from the past, the geographer often has recourse to memories and recollections of the place's past. These memories may be embedded in a monument or artifact of some kind. On the other hand, the past of places or the places of the past can still be present in the memories and recollections of individuals and social groups who have their places as a support for their lived experiences.

In another line of geographical thought, Professor Mauricio de Almeida Abreu (1998, p. 82), defines the term memory as the "capacity to store and preserve information", being "an essential element of a place's identity". By bringing together different individual memories, the memory of a place becomes a collective memory. In this sense, the subjectivity introjected into individual memory can converge into intersubjectivity, since different memories can converge to create a shared memory. We can then highlight the premise that the individual memory of the residents of a neighborhood, for example, can contribute to the recovery of their memorable geographies or their geographic memories (LOWENTHAL, 1982; 1998; ABREU, 1998).

Let's go on to appreciate the assumption that geography is everywhere and that each individual can be considered an informal geographer, able to discern the multifaceted universe they have experienced (LOWENTHAL, 1982; COSGROVE, 2004). Based on this premise, through the records contained in this individual's geographic memory, it is possible to walk the paths of their memories and uncover geographies of different temporalities, as well as past spatialities and artifacts that have been lost in the vortex of time, spatially speaking, but remain alive and pulsating in the memory of the human beings who experienced those moments. "The importance of this recovery for the identity of a place is unquestionable." In the midst of a country that tends to neglect its past, "oral histories (and geographies)", as well as the "memories of old people" (and of adults and young people, too), are in the process of being disseminated in Brazil (BOSI, 2003; ABREU, 1998, p. 83).

The experiences lived by individuals in a locality give rise to memories which, although very different from each other, have in common the adherence to the same place. Pursuing the necessary recovery of "time in place", or "the set of temporalities" (ABREU, 1998, p. 94) that makes it possible to unveil its past and present geographies (FERNANDES, 2010), we seek to elucidate the ideas related to the construction of a geography based on memory. After all, "it is impossible to deal with the empirical without coming to it with a previous theoretical background" (ABREU, 1998, p. 88). From these elucidations to others, having traveled the path so far with the purpose of philosophically grounding our approach, considering the verbal accounts of the residents of Ilha de Guratiba, we will now move on to the exercise concerning the fusion of accounts x philosophy. In the light of humanistic thinking, which does not distinguish the individual from their place, we will then use the geographic memory of the Guaratibanos to unveil an island of memorable geographies.

3 THE INVESTIGATIVE SCALE OF A NEIGHBORHOOD AFFECTIVELY CUT OUT

Ilha de Guaratiba is a place that can be used as an example of the city's expansion process. In the 1970s, the place, characterized as one of Rio's last rural remnants - a producer of fruit, vegetables and legumes - constituting the peri-urban fringe par excellence, was the stronghold of traditional farming and market families, as well as ordinary workers. From this period onwards, however, the notoriously bucolic place also began to be frequented sporadically by owners of second homes. Once this situation had been overcome, the place underwent a constant process of land/real estate appreciation, to the extent that it was home to a considerable increase in its resident population (FERNANDES, 2014 b). The decline of agricultural production turned Ilha de Guaratiba into an area of notorious real estate greed, as evidenced by the residential condominiums that sprang up after 1990. It is worth noting that most of the new residents migrated from neighborhoods in the South Zone and other locations such as Recreio, Tijuca and Jacarepaguà in search of closer contact with nature (FERNANDES, 2010; DIAS, 2011; LESSA, 2001).

To understand this picture of metamorphosis on Ilha de Guaratiba, let's highlight some aspects of the evolution of the city of Rio. In fact, since the first centuries of colonization, landfills have contributed to Rio's urban expansion. As the city grew, man overcame elevations by drilling tunnels, such as the Rua Alice/Barão de Petrópolis, in 1887, and the so-called Túnel Velho, in 1892, connecting Botafogo to Copacabana. More recently, Barra da Tijuca and Recreio dos Bandeirantes have gained expression amid the sprawl of Rio's urban fabric (CARVALHO, 2004), due to amenities such as the beach and other attributes.

In the current race to the west, the aforementioned urbanization march continues to incorporate new areas into its valued space. With a view to the 2016 Olympics, the Grota Funda Tunnel, inaugurated in 2012 as part of the Transoeste corridor, definitively integrates Ilha de Guaratiba into Rio's urban fabric (FERNANDES, 2014 b).

This summary represents a significant part of the approaches we will try to decode throughout this research, always starting from the memories shared by the residents of Ilha de Guaratiba. In this section, therefore, our goal is to introduce the real authors of this work: the people of Guaratiba. What do the residents of Ilha de Guaratiba think and feel about their "lived neighborhood"? Good! Let's leave the answers to these informal geographers.

3.1 The emblematic devotion of the co-authors of the research: the Guaratibanos

In order to understand a little more about the soul of the residents of Ilha de Guaratiba, let's examine the story of a Guaratiba native:

> A Guaratibano is someone who was born on Ilha de Guaratiba and has remained here. This is generally the case, since most of our residents have a history here, having accompanied the great changes that have taken place and those that continue to take place on our island. Guaratibanos are people who know everyone, who identify with local customs. The true Guaratibano is the one who has lived here for a long time and has deep roots in the place. This is my case. I was born here and have never moved away. My parents and grandparents are also Guaratibanos by birth, permanence and heart (Alex - electrician - 42 years old).

Electrician Alex's particular understanding of himself and his fellow citizens is illustrative, but relevant to understanding who the residents of Ilha de Guaratiba are. Armed with an unbridled passion for the place he lives, his opening statement emphasizes: "a Guaratibano is someone who was born and has remained on Ilha de Guaratiba". His statement can be analyzed when we consider the study by Dias (2011). In her findings on the continuous flow of new residents to the area, researcher Alice Ferreira Rodrigues Dias points out that after the arrival of new residents, two distinct groups of residents began to coexist on Ilha de Guaratiba: the "established" - residents of long descent with a family history related to agricultural work - which she confirms by this text stressed in previous paragraphs, and the "outsiders", these being the residents who chose Ilha de Guaratiba as their first or second home (ELIAS; SCOTSON, 2000; DIAS, 2011). "Outsiders" and "old residents" are another distinction that Dias (2011, p. 92) establishes between new residents and long-standing Guaratibans, such as the family of Alex, the electrician, who is rooted and tied to the place he lives in by memorable topophilic ties.

Regarding the Guaratibanos, let's take a look at what another "established" resident of the place tells us:

> Guaratibanos have the face of their place. They are very quiet people. They are members of traditional families, almost all related to family farming, which characterized the place, and to the world of street markets. My father is still a market trader. But unlike when I was a teenager, when he kept a vegetable garden at the back of our house, today he buys everything he sells at the market at CEASA. Guaratibanos usually live in a house with a big yard or on farms that now produce ornamental plants (Fabio - machine operator - 38 years old).

Like Alex, and using Yi Fu Tuan's thoughts on topophilia (1980; 2012), which refers to the affective link between the person and the place, machine operator Fàbio also refers to the long tradition of Guaratiba's families. Both emphasize the roots that are firmly planted in the place and the state of being well-liked by the community. In fact, the families of Ilha de Guaratiba are among the first in Rio de Janeiro, according to Carlos Rheingantz (1965). In his genealogical study, the researcher in question mentions several traditional Guaratiba families who already lived there between the 17th and 18th centuries. Among the families mentioned are the descendants of the "Abreu" family, honored with the name of a street in the town (Estrada do Abreu) and the descendants of the "Cubas" family. The latter descend from the patriarch Bràs Cubas, father of Gerônima Cubas, who arrived in the area after the constitution of the old "sesmaria de Guaratiba", founded in 1579, in the sixteenth century.

In view of Fàbio's statement that "the Guaratibano has the face of his place", we can underline Halbwachs' idea (1968; 2013), since the aforementioned thinker also understands that "the inhabitants resemble the neighborhood or the house. In every age there is a close relationship between the attitudes and spirit of a group and the aspects of the places in which they live" (HALBWACHS, 2013, p. 88). With this in mind, it's not difficult to understand the similarity between individuals and their lived world. When a place undergoes transformations of various kinds, these changes interfere in the relationships of its inhabitants. The opposite is also true, since changes in people's attitudes can also metamorphose their experienced ground.

Fàbio also mentions the traditional skills of the residents of Ilha de Guaratiba when the place was still devoted to food production. Until previous decades, a considerable portion of the local population was made up of farmers who, as a rule, also worked as market traders. This picture began to change when the decline of the street markets triggered a process of productive restructuring in the area. More recently, under the influence of Roberto Burle Marx's followers, Guaratibanos have become the largest producers of ornamental plants in the state of Rio de Janeiro.

In short, until the end of the 1980s, local production of fruit, vegetables and legumes served the outskirts of the city of Rio de Janeiro. Later, the production of ornamental plants replaced traditional agriculture, serving niche markets in all the states of the Southeast. This connection with the land, from which the inhabitants of Ilha de Guaratiba have always made their living, is the most emblematic characteristic of the long-established Guaratibano, contributing to the extraordinary affection these individuals have for their neighborhood.

3.2 **Ilha de Guaratiba: a neighborhood affectionately cut out by its residents**

For city managers, Ilha de Guaratiba does not have the status of a neighborhood in the city of Rio de Janeiro. But do its residents share this idea? Or do Guaratibans consider their place to be a neighborhood? Let's follow what Mr. Bento says about his "lived neighborhood":

Ilha de Guaratiba is my neighborhood. Our place is unlike any other in the city. Try to find a neighborhood in Rio de Janeiro that has similar characteristics to ours? To this day I don't understand why the city council doesn't recognize Ilha de Guaratiba as a neighborhood. When I fill in any form or am asked about the neighborhood where I live, I always answer that I live in Ilha de Guaratiba (Mr. Bento - market vendor - 67 years old).

As Mr. Bento points out in the previous lines, Ilha de Guaratiba has not yet been officially demarcated by the public authorities as a neighborhood. Just like Cinelândia: a name that has never been officially registered, but is enshrined in the hearts and minds and mouths of the people (www.roteirosdorio.com), Ilha de Guaratiba represents a neighborhood that Guaratibanos feel strongly about. Another example of neighborhoods chosen and affectionately cut out by their residents can be seen in Bandeira (1999). In her specialization monograph, the geographer studies Mallet, the name of an unofficial neighbourhood in the West Zone of Rio de Janeiro, officially incorporated into the neighbourhoods of Realengo and Magalhâes Bastos. Mallet has a predominantly middle-class population and is a residential neighborhood with vibrant commerce. Although it is not recognized by Rio's City Hall as a neighbourhood, like Ilha de Guaratiba, Mallet has been elevated to this level of affection by its residents.

On the subject of neighborhoods, geographer Marcelo José Lopes de Souza (1989, p. 150) understands that, consciously or unconsciously, people "always demarcate their neighborhoods based on landmarks" that they have established, "producing a symbolic heritage that they pass on from generation to generation". For Souza, neighborhood boundaries can even be imprecise or vary from person to person. This variation, argues Souza (1989), should not be too great, otherwise we will hardly be dealing with a neighborhood, since there will be no support for a Shared identity for its residents (SOUZA, 1989; 2003).

As we can see, a geographical neighborhood is not one that is demarcated by a political boundary that does not take into account its physical and human specificities. A real neighbourhood is one that is recognized, cut out and elected by its residents (SOUZA, 1989; HALLEY, 2014). "In the course of time", however, as Màrio Lago (1976) argued in his book, this perspective has had a major impact even within the city of Rio de Janeiro. A neighborhood like Tijuca had its boundaries extended by the greed of real estate speculators and, in this wave, residents from more or less distant neighborhoods like Praça da Bandeira, Estàcio, Rio Comprido, Andarai and, among others, even Vila Isabel and Grajaù. Nowadays, the loss of neighborhood ties, income and fear of the other have altered the boundaries of neighborhoods. In the case of Tijuca, status and income have led to tenuous and fluid boundaries. In fact, the boundaries vary from organization to organization and from person to person. The parishes established the most remote and traditional boundaries in the context of the city through or from the sesmarias and engenhos and their chapels. The City Hall and its IPTU uses another division. The Post Office has its own demarcation of neighborhoods. CEG uses another, and so on.

Next, let's take a look at the testimony of civil servant Marcos Sardinha about where he lives:

> The Grota Funda Tunnel has facilitated both access to our place and access to different parts of the city. For me, Ilha de Guaratiba is a strategic neighborhood in the West Zone. I believe that Ilha de Guaratiba is the converging point of the West Zone. Our neighborhood is close to Recreio dos Bandeirantes, Vargens (Vargem Grande and Vargem Pequena), Barra da Tijuca and Jacarepaguà - as well as Campo Grande and the surrounding area. Ilha de Guaratiba is to the West Zone as Méier is to the North Zone. In this sense, the tunnel only adds value to the neighborhood, since it makes it easier for us to get around and removes us from our historical isolation due to the Serra da Grota Funda, which divided the West Zone into two parts (Marcos Sardinha - 36 years old - civil servant).

In the words above, Marcos Sardinha not only considers Ilha de Guaratiba a neighborhood, but also elevates the place to the status of "converging point of the West Zone". This resident's opinion is obviously based on the strategic location of the place, established between neighborhoods in the Traditional West Zone (Santa Cruz, Campo Grande and surrounding areas) - located to the west of the elevations belonging to the Pedra Branca Massif - and the other neighborhoods in the coastal West Zone (Recreio dos Bandeirantes, Barra da Tijuca...), positioned to the southeast of the Serra Geral de Guaratiba (TARGINO and MONTEIRO, 2000). All this comes together to establish locational advantages endorsed by residents. The tunnel comes in the wake of the ambition to turn the area into a residential neighborhood as part of the Recreio de Guaratiba project, which has been advocated since the César Maia administration, integrating the Baixada de Guaratiba into the city's high-income coastal strip.

Figure 1 - Ilha de Guaratiba is part of the Guaratiba Administrative Region (26), located in the west of the municipality of Rio de Janeiro. As it is a neighbourhood not recognized by the municipality, there is no official map of this place.

Source: PROCESSAMENTO digital. Available at: <http://www.processamentodigital.com.br>. Accessed on: 20 nov. 2015

The official division of the city of Rio de Janeiro into neighborhoods only took place in 1981, by means of Decree No. 3158 of July 23, 1981 (Annex 1), based on the concept of the geographic notion of neighborhood defended by Soares, republished in 1990. This study was drawn up by geographer Maria Therezinha Segadas Soares in 1962, within the framework of classical geography (SOARES, 1962), based on the approach of French geographer Pierre Monbeig (1957), according to whom a city is a collection of neighborhoods, each of which has its own physiognomy, resulting from the function of its inhabitants and their age. All these neighborhoods, more or less integrated with each other, make up the city. For Monbeig, an urban neighborhood has its own unique character, a particular life, a soul. Influenced by this notion, Soares (1990, p. 105) lists:

the notion of neighborhood is of popular origin, taken from everyday language. For the inhabitant of a city, the neighborhood constitutes, within the city, a whole that has its own originality (...). The popular notion of neighborhood is much more geographic and richer, since it is based on a collective feeling of its inhabitants.

In zoning the city into neighborhoods, Soares (1962; 1990) considers the role of elevations to be important, since they can separate parts of the city, isolating them from the others. For the researcher, this physical characteristic can contribute to a place acquiring the individuality of a neighborhood. In this approach, the elements that could contribute to the individualization of a neighbourhood - in addition to the relief - would be the social content, the urban landscape and the function. The forces of the past and the factors of the present also play a part in shaping a neighborhood. Often, however, the municipal administration doesn't use this criterion, disregarding neighborhoods authentically elected by their residents, as is the case with Ilha de Guaratiba.

Again using the researcher Marcelo Lopes de Souza (1989), who looks at the concept of the neighborhood as a category of analysis, it cannot be defined by neglecting its individuality and the life of relationships that make it dynamic. In this sense, "any neighborhood is simultaneously an objective and subjective/intersubjective reality" (SOUZA, 1989, p. 148). In its phenomenological approach, "these two dimensions interpenetrate" (SOUZA, 1989, p. 148). Like any place, the neighborhood is lived and experienced by its residents, making them (residents and neighborhood) an inseparable entity (SCHUTZ, 1979; TUAN, 1980; 1983; 2012; 2013; MELLO, 1991; 2000; FERNANDES, 2010; 2014 a; HALLEY, 2014). In his proposal for a holistic vision, dialoguing with humanism in geography, Souza (1989, p. 149-150) points out:

The neighborhood has an identity that is intersubjectively accepted by its residents (...). A phenomenological look at the constitution of neighborhoods shows that the neighborhood corresponds to a certain part of the city which, by virtue of social relations, constitutes a lived and felt space for the individual. The recognition and feeling towards the neighborhood comes from the fact that it is the place where we were born, where our house is, where our friends' houses are, the square we frequent (...). This range of feelings is essential to a neighborhood's identity.

Although he points to intersubjectivity as an essential element in characterizing a neighbourhood, Souza (1989) does not ignore its objective dimension. For him, it is imperative that the two dimensions that make up the neighborhood interact in order to awaken in the residents a feeling of belonging to their neighborhood that the geographer, based on Tuan (1980), calls neighborhoodphilia (SOUZA, 1989, p. 151).

Regarding the place and lifestyle adopted by young Guaratibanos, let's take a look at student Marcos Wesley's account:

> People who live in Ilha de Guaratiba end up adopting the lifestyle that is common in the neighborhood. As I'm religious, I'm very busy with my church's activities. As well as weekly services, I have rehearsals for the theater group and the band I play in. Like me, most Guaratibanos belong to a church, either Catholic or Evangelical. I also like to work out and enjoy a good walk along our trails at the weekends (Marcos Wesley - student - 17 years old).

Like most of the residents of Ilha de Guaratiba, student Marcos Wesley enjoys a healthy lifestyle, in contact with nature. For this, he can count on the three large bodybuilding and gymnastics gyms in the town (Star Fit, W.M. Fitness and Ele e Ela) and, above all, on the roads, trails and natural viewpoints that characterize the place. Almost all the long perpendicular roads to the east of Estrada da Ilha converge at the foot of the Serra Geral de Guaratiba, an elevation belonging to the Pedra Branca Massif that surrounds the area. Walking along the Tachas road, for example, it is possible to skirt the less steep part of the elevation in question, reaching the town known as Rio da Prata, in the Campo Grande district. Another, much steeper trail can be taken up the Estrada do Morgado. This can lead directly to the Vargem Grande neighborhood, on the opposite side of the mountain.

Another expressive approach that Marcos Wesley (named after the founder of Methodism, the Englishman John Wesley) makes in relation to the place concerns the religious devotion of the Guaratibanos. The student whose story is highlighted declares himself an evangelical, and is a member of the Assembly of God Church in Ilha de Guaratiba. The Protestant denomination is one of many with temples in the neighborhood. They include the Church of Ilha de Guaratiba (Mount Hermon Baptist Church), the Maranatha International Ministry, the Projeto Atenda Evangelical Community, the Tabernàculo Restaurado Evangelical Community, among others.

Catholic temples also stand out on Ilha de Guaratiba. In addition to simple and beautiful chapels such as the one dedicated to Santo Sudàrio, located inside the site belonging to Doctor José Humberto Rezende (figure 2), a member of the Brazilian Academy of Fine Arts and a Santo Sudàrio scholar (FARAH, 2003), there are temples such as the Igreja da Matriz de Guaratiba (figure 3), the Igreja de Sâo Paulo Apóstolo and the Igreja dedicada a Santana, strategically located in Largo da Ilha de Guaratiba.

Figure 2 - Image from the advertising folder for Sitio Paraiso Verde, the secondary residence of Dr. José Humberto Rezende, where the Chapel of Santo Sudàrio is located. Due to criteria adopted by the site's owner, images of the chapel cannot be shown, but are restricted to visits by the faithful and researchers.

Source: SiTIO Roberto Burle Max. Available at: <http://sitioburlemarx.blogspot.com.br>. Accessed on: November 20, 2015.

Figure 3 - Guaratiba parish church

Source: KORYTOWSKI. Available at: <http://literaturaeriodejaneiro.blogspot.com.br/>. Accessed on: 20 Nov. 2015.

Regarding the religious phenomenon and its relationship with geography, Rosendahl (2002) points out that, as it is an integral part of human life, religion helps us to understand human feelings towards certain symbolic places, such as pilgrimage sites and temples. In another study, Professor Zeny Rosendahl (2003, p 205) states that:

> The body of work linked to the humanistic perspective of geography focuses on the connections that people develop with sacred places (...) contributing to the development of personal bonds and a sense of place. Tuan's research (1978 and 1980) states that an emotional bond is created and maintained through the construction of the sacred place. The process of creation contributes to places and objects becoming part of our self-identity, as does repeated contact, familiarity with the place and shared experience.

In fact, "the anchoring of identity phenomena in places" (BOSSÉ, 2013, p. 226) contributes to human beings also identifying with the place where their faith manifests itself. It is impossible not to attribute feelings of value to the locus of divine revelation (TUAN, 1978; ROSENDAHL, 2002; 2003; CASSIRER, 2004).

Another memorable geography of Ilha de Guaratiba comes to light through the testimony of resident José Maria. Let's see what the journalist shares:

> Ilha de Guaratiba has the ideal conditions to become a different neighborhood, an efficient neighborhood, following a trend that already exists in Europe and even in Brazil (Vila Madalena - São Paulo). We can no longer stand by and watch as the neighborhood and nature are destroyed, as has been happening all over the city. The most recent examples are right here on our doorstep: Barra and Recreio, which used to have phenomenal biodiversity and exceptional scenic beauty, with one of the most beautiful lagoons in the world, all wiped out by the sheer volume of construction and the consequent destruction of the environment and local quality of life. On Ilha de Guaratiba, if the residents remain as complacent as we all are, the phenomenon will repeat itself. Soon we will all be drowning in Coca Cola, Bobs and Mac Donald and the shopping malls. We're also going to see the destruction of the spectacular ornamental plant production culture, leveraged worldwide by our late Roberto Burle Marx. The time is approaching when the new PEU for the region will be discussed in public hearings with the community. It is necessary that everyone is mobilized, in their representative entities, discussing and studying the guidelines of the project already carried out by the City Hall's Urban Planning Department (José Maria - Journalist - 70 years old).

Influenced by the assumptions of the movement known as "localism", resident José Maria states above that Ilha de Guaratiba can become a self-sustaining neighborhood. Localism suggests that residents turn their lives towards the neighborhood itself. In this sense, people should try to work, consume and have fun where they live, boosting the local economy and preventing resources from flowing elsewhere (COSTA , 2013). With regard to this trend, Hardy (2013, p. 1) points out that:

> Leading urban planners, attuned to modernity and the enormous changes brought about by the technological age, maintain that the concept of sustainability no longer meets the dynamics of cities today. Its overuse in marketing campaigns and one-off actions has compromised the complexity of the issue. A new way of living with the future is now emerging. The efficient city/neighborhood. We can no longer allow urban growth, megacities and swollen neighborhoods, which are so prevalent all over the world, to continue. New planning is needed, in smaller units, with adequate population density, greenery and preserved local culture, good places to live, where the citizen's minimum needs such as health, education, work, security, housing and entertainment are met without the need for long commutes. Everything designed for people and not for cars.
>
> Mobility, currently one of the biggest problems in the big cities, will be made easier by avoiding huge expenditure on public transport, large hydroelectric projects, tunnels and viaducts. Stockholm, in Sweden, and Copenhagen, in Denmark, considered to be the cities with the best quality of life on the planet, are examples to be followed with the corrections appropriate to each region, the people and the technological world. Ilha de Guaratiba, with a low population density for historical reasons, a strong rural culture and a good quality of life, is the ideal place to start urban planning work to create a new model of urban life. We already have most of the necessary premises here. Ninety percent of the residents

work in the neighborhood producing ornamental plants, vegetables and fruit. Leisure is done right here in the neighborhood, in the clubs, on the soccer field, in the party houses. There is already a culture of cycling. In addition to all this, it should be noted that the city of Rio de Janeiro, according to the 1970 IBGE census, is the municipality in the state of Rio de Janeiro that has the largest number of people involved in agricultural activities, even beating Campos, which has just over 9,100 people in rural areas, while the municipality of Rio de Janeiro has 9,300 workers spread out in Santa Cruz, producing coconuts and cassava, and on the island of Guaratiba with ornamental plants. In order to preserve this valuable rural and cultural heritage, it will be necessary for the municipal, state and federal authorities and civil society to work together to create appropriate regulations for land occupation and use and to enrich the region with public facilities such as hospitals, schools and libraries.

When talking about the characteristics and attributes that, in his opinion, could make Ilha de Guaratiba an "efficient neighborhood", José Maria also mentions the Urban Structuring Project that the Guaratiba Rural Producers Association (Rural Guaratiba) submitted to serve as the basis for the new zoning that Rio's City Hall intends to implement in the area (appendix 2). José Maria, a rural plant producer, is in favor of a UDP that favors the maintenance of the local ornamental plant producers. The Association of Residents of Ilha de Guaratiba - AMIG - advocates an Urban Plan that maintains the natural attributes of the neighborhood, giving priority to environmental issues (annex 3). Amid the concern of rural producers and residents in general, the Guaratiba UIP - drawn up through the Institute of Architects of Brazil - has been sent to the urban planning department and could be approved at any time (appendix 4). In the context of the voracious real estate speculation that is transforming Ilha de Guaratiba into an urban site par excellence, we are seeing the neighborhood increasingly "flowing into Olympic Rio".

4 THE URBANIZATION MARCH ON ILHA DE GUARATIBA AS UNDERSTOOD BY ITS RESIDENTS

For humanistic geographers, it is important to emphasize the existential way people live in the places they live in or visit and, in this context, to draw experiences from them.

Despite this perspective, we believe that spatial metamorphoses are also responsible for a range of events that are relevant to the aura of the place, directly or indirectly influencing those who experience it.

In relation to the continuing spatial changes that have been taking place on Ilha de Guaratiba over the last few decades, we can see that, over the course of time, symbols of yesteryear and today - both material and immaterial - have been chosen, further strengthening the relationship between Guaratiba and its lived universe. When a place undergoes transformations, these changes can also produce existential metamorphoses, modifying people's relationships with their place. So we think: the different geographical contexts served as a backdrop for the experiences lived in the place. These should not be neglected, which is why we decided to explain them.

In this vein, this chapter discusses the urbanization march on Ilha de Guaratiba in the understanding of its residents - from its genesis to the present day - starting with its past rural-agricultural configuration.

4.1 The rural-agricultural context of yesteryear

The long geographical construction of Ilha de Guaratiba fills its residents with pride. The unbridled love for the place brings some of its residents back to its history. In this context, Tadeu Garrido emphasizes:

> Last week I was talking to a client who lives in Campo Grande about the common past of our places. On this occasion, he insisted that our locality was an offshoot of his neighborhood, which was much older than Ilha de Guaratiba. Nonsense! I told him that our place was 436 years old in 2015 and is one of the oldest places in the city (Tadeu Garrido - lawyer - 48 years old).

As in the words of the resident recorded above, Ilha de Guaratiba dates back to 1579, the year in which Manuel Veloso received from the Portuguese Crown a plot of land of approximately 52 km^2 to settle on. We are talking about the old sesmaria of Guaratiba, created years after the founding of the city of São Sebastiâo do Rio de Janeiro (FRIDMAN, 1999). This area, however, was inhabited by the Tupi Guarani Indians, probable descendants of the "pre-historic man of Guaratiba", who arrived thousands of years ago (MENEZES, ARAÙJO and GOES, 1998). The routine of the indigenous people only changed with the arrival of Manoel Velloso, who - together with his wife, Jerônima Cubas, daughter of Bràs Cubas - came to live in the newly constituted sesmaria of Guaratiba with the other members of his family and, from then on, began to build and manage sugar and brandy mills for export (PINTO, 1986). With regard to the activities of yesteryear, which were responsible for the prosperity of the old sesmaria of Guaratiba, let's consider the account of the university student Vitória:

> Studying and finding out about our roots from conversations with my grandparents, I fell in love with an unknown past. So I went to study, to find out what Ilha de Guaratiba was all about and, I confess, I'm very proud of what I discovered, as this was a prosperous area in the past. There were many farms in the area and sugar mills attached to them. The "Alambique dos Mudinhos" - which was part of my childhood - was one of these mills in the past (Vitória - student - 22 years old).

In the second half of the 16th century and the following, Ilha de Guaratiba was a large rural estate where sugar mills, cattle and horse farms and fruit, vegetable and fruit plantations dominated. At that time, it expanded rapidly, thanks to the hard work of Velloso's descendants. By the end of the 18th century, many sugar mills were multiplying, including Engenho Novo, Engenho de Guaratiba, Engenho da Ilha and Engenho do Morgado, which once belonged to Father Joâo Pereira de Cerqueira and is now the Mudinhos still, as resident Vitória pointed out. In order to transport the produce, the mangrove swamp had to be cleared in order to support this demand. In Pinto (1986) there are reports of some boats being shipwrecked on the river, when the sugar production from Engenho do Morgado was being transported. The sugarcane plantations and sugar mills were joined, from the 19th century onwards, by coffee plantations, cultivated on the hills of Ilha de Guaratiba (FRIDMAN, 1999).

Over the centuries, sugarcane and coffee plantations, as well as the other rural activities mentioned above, left their mark on the area, which characterized Ilha de Guaratiba until the second half of the 20th century. The great agricultural production of yesteryear made Guaratiba one of the richest and most prosperous parishes in old Rio in the 18th and 19th centuries, until a great drought in 1888 consumed its crops (SANTOS, 1965; FRIDMAN, 1999). In the context of the outskirts of the city, until the first half of the last century, Ilha de Guaratiba represented the so-called "sertâo carioca", and its ruralism and rusticity was the focus of an important study by researcher Magalhâes Corrêa in 1936 (CORRÊA, 1936). Because they live in a peripheral location, far from the central area, the residents of Ilha de Guaratiba are often the target of jokes and mockery from those who live in better-known neighborhoods. Let's take a look at the following story:

> Back in college, when I said I lived on Ilha de Guaratiba, people would ask: where is this place? To this day, when I tell people at work where I live, they are perplexed, as if I lived in another dimension. Occasionally, my coworkers don't miss the opportunity to make jokes like: "You'd better go now if you don't miss the 10 o'clock vine" or questions like: "Has the electricity arrived there yet?". I take it all in jest, but deep down I feel that they think I live badly. However, I never miss an opportunity to tell anyone that I wouldn't swap this "end of the world" for anywhere else (Marquinhos - 30 years old - civil servant).

For many who hear about Ilha de Guaratiba, it is nothing more than a peripheral area far from everything and everyone. However, since their passion for life doesn't share their dulled thinking, in an ethnocentric attitude towards geography, Guaratibanos like Marquinhos place their place at the center of their world. Representing a common human tendency, ethnocentrism leads people to overvalue their lived universe, elevating it to the status of the best place in the world. This is the positive side of ethnocentrism. However, this attitude also has negative aspects. In this sense, people who live in the city's upscale neighborhoods tend to discriminate against those who live in areas they consider less important (TUAN, 1986; 2013; MELLO, 1991).

Even with the denomination of "distant periphery" and representing the so-called "rural area of the city", Ilha de Guaratiba - between the 1940s and 1960s - was already considered by the government to be part of the "intermediate periphery of the metropolitan area" (ABREU, 2008). In relation to the temporalities described in this paragraph, let's see what a Guaratiba woman with a long tradition says about the geographic context of her life:

> Ilha de Guaratiba used to be a place where vegetables, fruit and legumes were grown. There were large banana plantations in the highlands and huge orange groves in the lowlands. My father was a big orange grower. Two large consignments of the fruit were harvested every year for export to the United States. When the war broke out, my father no longer had anyone to sell to and the oranges were no longer exported. At that time (1940s), my father began to replace part of his orange grove with vegetables and other agricultural crops (Mrs. Antonieta - 86 years old - housewife).

Like a considerable proportion of Guaratibanos with a long tradition, Dona Antonieta is the daughter of Portuguese parents. Her parents arrived in the town in 1925, a period when access to sizable plots of land was still relatively easy due to the human vacuum that the area represented at the time. In the meantime, citrus farming was practiced by most of the large agricultural producers in the area. Having just arrived from Portugal, Mrs. Antonieta's father had no hesitation in deciding to produce oranges, since the large American consumer market was a guarantee of the large production. One of Magalhães Corrêa's descriptions of the area in question during this period is "the region of the orange groves" (CORRÊA, 1936). To further qualify Mrs. Antonieta's words regarding her statement about the orange groves, let's freely consider Abreu's (2008) elucidations. The researcher points out that citrus production acted as a major brake on the wave of land development until the first half of the last century, preventing some municipalities in the Baixada Fluminense and neighborhoods in the traditional West Zone of Rio de Janeiro from being hit by the real estate fever of the time. However, with the outbreak of the world conflict, exports collapsed, as all the orange was exported on foreign refrigerated ships that no longer docked in Rio de Janeiro. In addition, the lack of cold stores and poor transportation by road and rail led to the fruit rotting on the trees, causing a citrus plague that decimated a large part of the plantations. At the end of the war, with production no longer serving the domestic market, the export of oranges was banned, dealing the coup de grace to those who had managed to preserve their orange groves during the crisis. From then on, the orange groves were replaced by housing developments in municipalities like Nova Iguaçù and neighborhoods like Campo Grande.

This urbanizing march, however, did not take place on Ilha de Guaratiba, since many of its orange groves were kept to meet internal demand and the rest were replaced by a diversification of agricultural crops that came to characterize the place as a true green belt - a major producer of fruit and vegetables. Still on the subject of the rural-agricultural context of the island of yesteryear, the following account is worth mentioning:

> The island of yesteryear, when people still lived off agriculture, was fantastic. I remember how, in the 1980s, a hang-glider flew over the massif and crashed in the middle of Manoel Fonseca's farm. At the crash site, the free-flyer told me that, during the flight, a strong gust of wind lifted him above the clouds and he was completely overcast for a few minutes. When he finally managed to lose altitude and regain his vision, he was faced with a landscape so different from his reality that, in his own words, he thought he had found the "missing link" (Jailton or "Pipo" - 49 years old - biologist).

The rusticity and rurality of the "sertâo carioca", with its inland aspect and its farmlands, already impressed Rio de Janeiro's urban "outsiders" in the 1930s (CORRÊA, 1936). Can we imagine this fascination 50 years later, when the city had already taken on the airs of a metropolis? It was surprising to most cariocas that, within their own city, in a metropolis like Rio de Janeiro, there was a place completely different from the world they lived in. A place characterized by farmland and a life devoted to nature and rustic countryside.

4.2 The decline of street markets and the dawn of landscaping

The agricultural activity in Ilha de Guaratiba, outlined in the previous topic, was closely related to the world of street markets, as we can see in the statement below:

> For decades, agriculture on Ilha de Guaratiba was maintained thanks to the street markets. Until the 1980s, most Guaratibanos had two jobs: farmers and market traders. When CEASA was built, there was a direct impact on local agricultural production. Many of the farmers who grew the products they sold at the street markets started buying them at CEASA. Then the local sacolões broke out, directly affecting the street markets. Most people stopped going to the markets when sacolões opened in their neighborhood. In addition, the prices at the sacolões were more attractive. I remember with nostalgia the great market traders and producers of the locality: Mr. Nori, a great producer of bananas; Mr. Manoel Fonseca, who produced vegetables; Mr. Alfredo, a great producer of chayote and passion fruit. Another factor that contributed to the ruin of the farmers of Ilha de Guaratiba was the competition represented by large producers from the Serrana region and the interior of São Paulo. Gradually, the local farmers realized that there was no way they could compete with the products coming from outside. Many of them replaced their fields with the production of ornamental plants. In addition, by becoming a supply center that receives products from various municipalities in the state, CEASA began to compete with the rural producers of Ilha de Guaratiba, since many market vendors began to buy products in this market that, in the context of the city of Rio de Janeiro, were only found in Ilha de Guaratiba. By competing internally with food cultivation from 1990 onwards, the production of ornamental plants - because it represents a much more lucrative activity - also contributed to the decline of the traditional agricultural activity on Ilha de Guaratiba (Jocenir Pestana - mechanic - 54 years old).

According to Mascarenhas (1991, p. 1), "the street market consists of a periodic group of retailers" who display their merchandise in versatile structures, using public roads, and therefore depend on concessions from the municipality to temporarily take over the streets (p. 12). Therefore,

> the market vendor does not own the space he uses, unlike traditional commerce. They only acquire, on a temporary basis, the right to use that space on days of the week and at times pre-established by the government, to display certain products (MACARENHAS, 1991, p. 13).

Widely dispersed throughout the city of Rio de Janeiro, the street market has played an important role in urban supply over time, especially in the food sector (vegetables, fruit and fish). From the elite neighborhoods of the South Zone to the suburbs of the West Zone, this periodic gathering is part of everyday social life in Rio. From the 1970s onwards, however, a new type of retailing came onto the scene: supermarkets. Their rapid expansion in the city ushered in a period of strong competition with the traditional periodical markets, compromising their performance and radically changing their spatial distribution (MASCARENHAS, 1991).

Looking at Jocenir Pestana's account of the past spatiality of Ilha de Guaratiba, we can see that the aura of the place favored agriculture and that most of the arable land was used by its owners - notably small and medium-sized farmers - to produce fruit, vegetables and legumes. These products were largely sold at street markets located in different neighborhoods. Local agriculture was geared towards this retail trade, providing Ilha de Guaratiba's producers with a market for their produce and supporting this activity for many decades.

For Jocenir, the breakdown of the agricultural system that had always characterized Ilha de Guaratiba was due to a series of events, including the gradual decline of the street markets. For him, the genesis and subsequent proliferation of the sacolôes in the city's neighborhoods dealt a severe blow to the street markets and, by extension, to the local farmers who depended on the dynamism of this retail market. The mechanic also mentions that the opening of the fruit and vegetable supply center (CEASA) in the Irajà district had a negative impact on local agriculture. For him, many market traders opted to buy directly from CEASA and gave up their fields. Jocenir, a mechanic, adds that, after its inauguration (in 1974), CEASA became a direct competitor of the rural producers of Ilha de Guaratiba in relation to the wholesale trade.

The decline of free markets, according to Mascarenhas (1991), began with the advent of supermarkets in the 1970s. This decline was aggravated by the proliferation of popular grocery stores (sacolòes) throughout the city from 1990 onwards (FERNANDES, 2006), leaving the majority of agricultural producers in

Ilha de Guaratiba without a consumer market for their products. The loss of dynamism of the open-air markets meant that traditional agricultural activity collapsed dramatically in the area and forced it to be restructured.

Despite the transformations brought about by the decline of its traditional agricultural activity, some traces of other times still remain in the area. An example of this are the farmers who, to this day, try to resist by continuing with their traditional vegetable crops. Other producers, however, have started to grow ornamental plants on the land previously used for agriculture, as resident Jocenir pointed out at the end of his statement. For him, the production of ornamental plants competed internally with agriculture, since it became a much more profitable activity than food production from 1990 onwards. Jocenir's account allows us to uncover part of the geographic context of the place, related to agriculture and its relationship with the street markets. This long-time Guaratibano also points to the dawn of new activities linked to the production of ornamental plants, gardening and landscaping, branches that have replaced traditional agricultural production.

In order to understand the process of transition from traditional agriculture to landscaping, I believe it is pertinent to look at one more statement:

> Before replacing agricultural production in 1990, landscaping coexisted with food production for more than 40 years on Ilha de Guaratiba. In the first three decades of coexistence, however, there was no competition between the two activities, since the free markets were still dynamic enough to guarantee the maintenance of the crops. However, with the emptying of the street markets from the 1980s onwards, many plots began to be replaced by ornamental farms. This transition was only possible thanks to the influence of the late Roberto Burle Marx, the forerunner of the landscaping activity that is prevalent on Ilha de Guaratiba today. Ever since he bought his place in 1949, Burle Marx - gradually - influenced other people to change their business. The largest garden in the area (Horto Rio Verde) was already developing its landscaping and gardening activities in 1973. Today, many of the agricultural producers of the past (Manoel português, Max, Mr. Alfredo...) have started producing ornamental plants on their properties, a much more lucrative activity. Until the 1980s, Ilha de Guaratiba was considered the city's great green belt, characterized by its vegetable gardens and orange groves. Today, we represent the city's largest plant production center, one of the largest in the state (Luis Carlos - trader - 60 years old).

According to Luis Carlos, Burle Marx's influence was decisive in the transition between agriculture and landscaping on Ilha de Guaratiba. A local resident, Burle Marx was struck down by abdominal cancer in June 1994, at the age of

84. When he died, he left behind 2,000 designed gardens. Born in São Paulo and moved with his family to Rio in 1913, he grew up in an environment surrounded by greenery in Leme, where from an early age he helped his mother grow plants in the family garden. A student of architecture and painting, the future landscaper went on to become a plastic artist, until his neighbor Lúcio Costa, who admired the plants he cultivated, invited him to do the landscaping for one of his projects. This is how, in 1932, Roberto Burle Marx signed his first garden - for the Schwartz family home - in Copacabana, making the 23-year-old famous. His last work was a project for Kuala Lumpur, Malaysia, which was on his drawing board when he died and was completed by his team (CALS, 1995; SA, 2008).

In order to cope with the large number of plants that his many gardens required, in 1949 he bought the former Sitio Santo Antônio da Bica, which today houses the Sitio Roberto Burle Marx. It was in this area of more than 35,000 square meters, located at the foot of the Serra Geral de Guaratiba, that Burle Marx began to produce, decorate and collect hundreds of ornamental species. Due to the many friendships he made and the great affinity he had with the place, as well as the need to be closer to his work, Burle Marx moved permanently to Guaratiba in 1973, thus intensifying his influence on the place (CALS, 1995; SA, 2008).

In the same way that fruit and vegetable production replaced citrus farming after its crisis in the 1940s and 1950s, guaranteeing the agricultural skills of Guaratibanos, floriculture - especially after 1990 - has been replacing the traditional gardens of Ilha de Guaratiba, as Luis Carlos pointed out in his statement. This transition, however, was only possible due to the prestige and influence of Roberto Burle Marx.

Being passionate about nature and the bucolic nature of its landscapes, Burle Marx - also a painter - chose Ilha de Guaratiba to live, work and produce his canvases, using its natural scenery as a backdrop and inspiration. This is how this eclectic artist produced hundreds of paintings, magnetized by the scenic beauty of the place he chose as an anchor and also to live his experiences (SA,

2008; TUAN, 2012; 2013).

Despite maintaining the basic activity of the land, the production of ornamental plants has been compromised by a new dynamic that is emerging in the area, related to the real estate market. Currently, the plants produced in the area are destined for the growing demand of condominiums and construction companies, whose "fetish landscaping" is one of their strategies for transforming land into a real estate commodity. In Ilha de Guaratiba, however, this speculative process had its milestone in the 1970s, when the first outsiders began to acquire plots of land that gave rise to the first second homes in the area.

4.3 **Secondary residences as the embryo of the urbanization march**

Another approach to Ilha de Guaratiba is contained in the following account, which is the gestation of the real estate speculation that began in the 1970s, making the place one of the most sought-after by those who wanted to momentarily escape the metropolitan bustle, even though they lived there. Let's take a look:

> I remember some families attending weekend masses in the 1970s. At that time, there were already many "summer" sites in the area. Most people, however, didn't come permanently. They bought a plot of land, built a nice house, and came here for vacations, vacations and weekends. Today, most of them live there permanently. The families of my neighbors - Dr. Osvaldo, Jorge Celso, Mr. Franklin, among others - arrived here in the 1970s. Some small landowners started selling their plots to these people from abroad. At first, they only came here at weekends. But then they all moved permanently to their respective sites (Benedito or "Dito" - Horto Rio Verde employee - 50 years old).

According to the resident above, the owners of second homes found in Ilha de Guaratiba what they were looking for: a place where the natural attributes and country atmosphere, associated with bucolicism and tranquillity, made it possible to escape from the metropolis. In contrast to the urbanity inherent in the neighborhoods where they lived permanently, Ilha de Guaratiba came to represent a kind of "escapist Eden" (TUAN, 1998) for outsiders who could afford to opt for a lifestyle focused on nature and the countryside. In this new "environment" (TUAN, 2012), the city dweller could take refuge from time to time with his family in order to have his strength renewed for the stressful metropolitan work. In an intellectual effort to define escapism, Mello (2001), based on Tuan (1998), points out that there are "various escapist levels", among which we can mention the natural and/or artificial paradises in which human beings seek a momentary escape from harsh reality. "When surrounded by problems or in unhappy moments, the individual unwinds, walking through other worlds" (MELLO, 2001, p. 97). In the midst of the stresses common to the urban environment, families from different parts of the city have tried to "escape" from these problems by moving to Ilha de Guaratiba.

In his elucidations on the inexorable process of the urbanization of society, a phenomenon that has been heralded for decades by various researchers (BELL, 1977; CASTELLS, 2002; CANCLINI, 1999; SANTOS, 2001; 2002; CAPEL, 2003; HARVEY, 1989; 2005), Lefebvre (2008) points out that a second home, in the middle of the countryside, can represent the embryo of an urbanizing march in this spatiality, since it is directly connected to the urban.

If we relate the phenomenon of the acquisition of second homes to an analysis of the urbanization process in Rio de Janeiro, we can't ignore the role of speculative real estate capital in the housing production process, which is one of the main factors responsible for the sprawl of Rio's urban fabric (ABREU, 2008; RIBEIRO, 1997). In this context, Ribeiro and Coelho (2007) point to the emergence of new forms of housing aimed by real estate capital at the most affluent classes, which have contributed to the spatial restructuring and expansion of metropolises. Residential condominiums are one example of this. However, this logic related to the production of urban space often starts with the acquisition of secondary residences by the economically privileged classes, as suggested by Assis (2003) for a general context and Fernandes (2003; 2006; 2009) for a specific context, based on the urbanization process of Iha de Guaratiba.

In his research on the expansion of the second home phenomenon, geographer Lenilton Francisco de Assis seeks to elucidate the main causes of this event, which coincides with our approach. In his analysis of the spread of the phenomenon of second homes, Assis (2003) points out that, as a result of the process of metropolization of certain cities, it became increasingly necessary for urban man to leave the overcrowded central areas in the direction of the metropolitan peripheries in search of a reunion with nature. This was a way of relieving their daily stresses and renewing their energy. The urban environment, which was once the center of attraction for housing and rural people in search of work, now, despite playing a variety of roles, leads its residents to seek out new areas that offer them the conditions they need to spend their free time in contact with nature. Therefore, due to their proximity to central areas, the metropolitan peripheries have become one of the targets of real estate speculators who seek to valorize the natural and cultural attributes of these areas, offering them to specific social segments who have the surplus income to buy a second home.

Assis (2003) also proposes that the phenomenon of second homes is one of the factors responsible for the urbanization of the periphery, since certain capital migrates to peripheral areas, materializing through properties that also come to represent a reserve of real estate value. By assuming the availability of surplus income, second homes are no longer just an alternative for leisure, but also an investment option. From the moment that the second home also has an exchange value, the person most responsible for transforming the peri-urban space comes into play: the real estate speculator. The latter, through insidious advertising, aims to transform the natural attributes and amenities of the metropolitan periphery into real residential attractions, as we can see below in relation to real estate speculation in Ilha de Guaratiba:

> Finally, it's your turn to live in paradise! Right here on Earth. Prawns and squirrels scurry across the asphalt. Oxen walk leisurely in single file towards the pastures. Marmosets in droves monkey around in the branches of the trees. Herons and wild ducks, in their lightness, develop scenic choreographies over lakes and streams in spectacular flights. That's life around here. Time seems to have stopped.
>
> It's definitely the new Recreio! The difference is that here you can still negotiate with decent country bumpkins at low prices. Mini sitios for you to live near the good and the best, "far and near" expensive and saturated areas (FERNANDES, 2006, p.42).

In the quote above, taken from an advertising folder (J. Brandao Negócios Imobiliarios), what strikes us most is the emphasis placed on the natural attributes of the place, where it is sold as a true paradise, lost in the middle of the metropolis of Rio de Janeiro. This real estate advertisement, broadcast in the early 1980s, reveals the genesis of a practice that is very common today, but with a very significant difference: the romanticism that the ad tries to convey is justified by the situation of the place at the time. In reality, at that time, Ilha de Guaratiba was nothing more than a town characterized by rural activities - mainly based on agriculture and the many farms visited by its owners during their leisure time. It was the typical weekend in the countryside. As such, this type of advertisement was aimed at people interested in acquiring a relatively large area where they could build a sitio which, in most cases, also housed a secondary dwelling. However, from the early 1990s onwards, new real estate agencies began to offer economically privileged strata a different product from that offered by their precursors. Secondary housing was replaced by residential condominiums, where housing was no longer temporary.

In order to confirm Benedito's statement when he explained how the second homes on Ilha de Guaratiba became permanent dwellings, let's return to the approach of Assis (2003). When looking at the change from second homes to permanent dwellings, the geographer explains that in the last phase of this process, the original perimeter is absorbed by metropolitan expansion, now forming part of the city itself, while the previous second homes are metamorphosed into permanent dwellings. In Ilha de Guaratiba, more and more people are choosing to live in their former second homes. In addition, families, mainly from the Recreio-Barra-Zona Sul axis, have found favorable conditions in the area, considering its locational advantages, such as relatively low prices, availability of land, landscape attributes, low crime rates, among others. The latter, however, represent a new type of resident: those who opt for horizontal condominiums.

Despite symbolizing a new spatial dynamic in Ilha de Guaratiba, the phenomenon of housing production - on a broader scale of analysis - reveals, as Ribeiro (1997, p.199) points out, the continuity of the process of urban expansion in the city of Rio de Janeiro, since:

> From the second half of the 19th century, especially after 1870, the city of Rio de Janeiro underwent important urban transformations generated by the action of a group of capitals that began to invest in

urban space. Among them, what we might call real estate capital, applied to the production of rental housing and the purchase, subdivision and sale of plots of land previously used for agricultural purposes...

The observation of Jocenir, Luis Carlos and Benedito's testimonies, outlined above, leads us to understand that the disarticulation of the rural-agricultural system of Ilha de Guaratiba, prior to the process of consolidating the place as a reserve of real estate value, triggered a process of transformation that has radically changed the locality. Considering that each place is a combination of temporalities and "geographic testimonies" (www.roteirosdorio.com), we can see in Ilha de Guaratiba the existence of elements representative of phases prior to the urbanization march in question that still resist in the midst of innovations. As examples, we can cite some remnants of rural activities from the past and the persistence of many sites that are still used as a leisure option during vacations and weekends (second homes). Since events such as the urbanization march in question represent continuous actions that imply time, continuity and change (SANTOS, 1992), we understand that the phenomenon of second homes represented the genesis of the metamorphosis that Ilha de Guaratiba has been going through. Following the flow of second homes (1970s/1980s), the first residential condominiums began to appear (from 1990 onwards), as a result of the growing real estate speculation in the area.

4.4 Real estate speculation and the advent of condominiums

The reasons for the rise in property values, which led to the first condominiums in the area and a growing number of new residents, are highlighted below:

Due to the tranquillity of the place, ideal for living and raising children, the first arrivals spread the word to their relatives and friends. These, in turn, were the ones who later arrived to live in the condominiums. As Barra da Tijuca and Recreio dos Bandeirantes grew and expanded, people who liked a more rustic lifestyle, focused on contemplating nature, rurality and tranquillity, began to come to Ilha de Guaratiba. When they found the place, they started spreading the word to their relatives and friends who had also come here through word of mouth. The first ones came to the ranches. The latter, to the condominiums. Today, people want to get away from the city, or rather from its problems such as pollution, violence, traffic, etc. They look for places like Ilha de Guaratiba, which, despite being part of the city, still doesn't share its common problems. The tranquillity and daily contact with nature allow the residents of Ilha de Guaratiba to live a quality life. Despite the growth, it is still possible to breathe the air of a country town. This is wonderful. The lack of space in the areas closest to the center and the flight to a quieter place, which can offer a better quality of life, are also factors that have contributed to real estate speculation and the consequent construction of the first condominiums on Ilha de Guaratiba. The wide availability of land, coupled with the relatively lower prices, as well as the proximity of Recreio and Barra, have contributed to many people coming to Ilha de Guaratiba (Sandra - 57 years old - teacher).

As a resident with a deep knowledge of her place, Sandra shares in her account some of the possible reasons for the real estate development that contributed to the construction of the current condominiums in the town. Natural attributes, the availability of land, the relative proximity of the central area and of neighborhoods with sub-centers of commerce and services, among others, are mentioned as the main factors of real estate development that promoted both the aforementioned residential mobility and the consequent genesis of the condominiums that were built to cater for this new and growing public. What's more, the tranquillity and countryside atmosphere of the place, as opposed to the urbanity and hustle and bustle of the city, are cited as a stimulus for migration to the west.

Created above all in the new areas of the metropolis with natural amenities, the exclusive horizontal condominium is the result of a process of effective land appreciation and real estate development. Becoming the "eldorado" for an upper middle class that comes, in part, from the city's former noble areas, condominiums are characterized by the self-segregation of social groups who, having an income, can live wherever they want. And the choice of new home is influenced by the massive advertising surrounding the amenities and the new lifestyle (CORRÊA, 1992).

In the 1990s, the first residential condominiums sprang up on Ilha de Guaratiba as a result of rising property prices. The area, which for centuries had been little explored and which locals said had stood still in relation to other places in the municipality of Rio de Janeiro, now began to attract the attention of a large number of people, in the midst of numerous metropolitan problems. Its domains began to be an object of desire when it came to living in a place full of amenities, offering those wishing to escape the urban neighborhoods a quality of life that matched their desires. In this sense, it is worth repeating that the attributes of the place, such as mountains, greenery, tranquillity and security, among others, are used as a lure for those who wish to be, paradoxically, close to and far from urban life (TUAN, 2012; FERNANDES, 2010).

As explained above, the genesis of the race towards the so-called "lost paradise" in the interior of the metropolis occurred when several second-home owners began to live there permanently. This phenomenon caught the attention of some real estate speculators who began to exploit the charms of the location in their advertising. This type of speculative activism persisted throughout the first half of the 1990s, thus increasing the demand for real estate on a scale that the old properties, such as sitios and mini sitios, could no longer cope with. The first horizontal residential condominiums then began to appear, built both on the slopes of the hills and on the flat areas of the site, especially on the tidal flats, through intense landfilling of mangroves and alluvial channels.

Since the first condominiums were built in the area, real estate prices have been rising. Every day it becomes more expensive and difficult to live in the area, especially in the condominiums which, with very few exceptions, are an option only for people from high-value areas such as Tijuca, Recreio dos Bandeirantes, Barra da Tijuca and neighborhoods in the South Zone. This is due to the high value placed on land, which is only accessible to economically privileged classes who, for the most part, live in the city's upmarket areas.

As a type of development that requires a great deal of land and is built on areas of environmental preservation, horizontal residential condominiums, as well as representing the pillar of a new urban-residential trend, are the most responsible for the reproduction of countless environmental impacts. Since the infrastructure that could provide better support for the sprawl of Rio over the area studied is non-existent, environmental degradation has become a constant concern for many Guaratibanos.

4.5 **Today's urban trend and environmental impacts**

In the wake of the urbanization march, nature gradually gave way to real estate developments, promoting a series of environmental impacts, a problem rejected by residents such as agronomist Ingo (53) in his statement transcribed below:

Urban dwellers love greenery. But when they come to a place like ours, they cut down the trees, pollute the rivers and end up turning nature into a city. Instead of greenery, trees and rivers, they build houses, walls, roads, etc. Many call this progress. For me, however, it's a regression. A rich place is one with clean water, preserved forests and clean air. That is the greatest wealth a man can have. If the current pace is maintained, the destruction of Ilha de Guaratiba will happen in the same way as Barra da Tijuca. The most beautiful thing about Barra was nature, which has given way to buildings, highways and shopping malls. Guaratiba Island **is the green lung of Rio de Janeiro**. This area should be preserved in its natural attributes and organic farming should be practiced on the slopes, within the forest. This should be **the** locality's **geographic identity**. Ecologically speaking, the process of urbanization is very problematic. Ilha de Guaratiba doesn't have lagoons like Barra da Tijuca and Jacarepaguâ. In 50 years, Barra has managed to clean up the waste that has been channeled into the lagoons all this time. Good or bad, the waste was purified. Because we don't have lagoons, the waste produced here tends to contaminate the rivers and the water table, from which a large part of the population draws its drinking water. The shrimp and frogs have disappeared from the streams. This is because tadpoles need crystal-clear water to develop. The freshwater shrimp, too, can only be found in clean water. The absence of frogs and shrimps in almost all the streams on Ilha de Guaratiba shows that these streams are polluted to drastic levels. I miss the symphony of the frogs every night and fishing for shrimp in the rivers that have now become sewage dumps (Ingo - agricultural engineer - 53 years old).

Ingo's account shows his concern about the impacts on the nature of the place caused by the implementation of new real estate developments and the increase in population. In his statement, Ingo makes clear his concern about how the new city dweller relates to nature and emphasizes his aversion to the urban. The agronomist also shows concern about the environmental risks arising from the impacts produced by this increase in the number of residents.

With regard to the feelings and sensations of individuals and social groups in relation to an attack on their lived world, exemplified by the apprehension shown by the resident in question in relation to the attack on their common environment, we can deduce, based on Tuan (2012), that environmental problems are fundamentally human problems. The aforementioned thinker is certainly based on the totalizing vision, also shared by Corrêa (1992), which includes the human element in its conceptualization of the environment. However, both in the midst of these introjections and in relation to the approach of human geography that privileges man, including him in its analysis, it is also undeniable that every human action on nature (re)produces environmental impacts (DREW, 2002).

Expanding the scale of analysis around the relationship between urban growth and the impacts caused to nature, it is interesting to refer to the emblematic process of consolidation of the metropolis of Rio de Janeiro which, compressed between the sea and the mountains, flanked by beaches, sandbanks, marshy lowlands and forests, had its growth forged in the struggle for space and the overcoming of distances, generated by this same growth, in the face of the special conditions of its physical environment (GALVÀO, 1992).

> Squeezed between the mountains and the sea, Rio de Janeiro had these natural elements as the great beacons of its expansion (...). The development of technology allowed these obstacles to be gradually overcome, enabling the city to incorporate into its built fabric spaces that were once considered unsuitable or unlikely for urban occupation (ABREU, 1992, p.54).

In addition to the massifs and mountains mentioned by Abreu, the spaces considered unsuitable or unlikely for the sprawl of Rio's urban network also included alluvial channels, sandbank areas, mangroves and forests. All these old limiting factors, however, were no obstacle to metropolitan growth. Thus, from the first centuries of colonization, landfills contributed to the organization of Rio's urban space. With the growth of the city, the elevations had to be broken up by drilling tunnels since 1887 (CARVALHO, 2002). More recently, Barra da Tijuca and Recreio dos Bandeirantes have gained expression amid the sprawl of Rio's urban fabric. In recent decades, Rio has grown towards the Guaratiba plateau (FERNANDES, 2006; LESSA, 2001), amidst the fears of its residents about the consequent deterioration of the natural biosystems that characterize the area.

Ingo's concerns, representing the community of this place, are justified because, as a lowland area with ample and exuberant greenery, surrounded in large part by the Pedra Branca Massif (Serra Geral de Guaratiba), Ilha de Guaratiba's most important feature is its beautiful natural landscape, which is the backdrop used by real estate agents to attract followers of a lifestyle based on closer contact with nature. The area is basically made up of two conservation units, namely the Guaratiba Biological and Archaeological Reserve, located on the tidal plain, and the Pedra Branca State Park, of which the Serra Geral de Guaratiba hills are also part (ATLAS DAS UNIDADES DE CONSERVAÇÃO DA NATUREZA DO ESTADO DO RIO DE JANEIRO, 1990). Furthermore, according to the Rural Union and Rio de Janeiro City Hall, Ilha de Guaratiba is an important area of rural and environmental preservation. As mentioned, this information is ratified by the city council itself, in a municipality considered to be urban.

Geographically speaking, all human activity advocates environmental transformations, since the environment - or space - presupposes, in addition to natural and built material forms, also the human being who gives it dynamism (SANTOS, 2002). Therefore, the more unprepared an area is in terms of basic infrastructure to support the desired changes, the greater the environmental impacts will be. This is the case of Ilha de Guaratiba. As if the changes represented by the influx of new residents and the construction of housing were not enough, the lack of infrastructure in the area further increases the damage to nature.

Considering only the actions promoted by the phenomenon described by Abreu (1992; 2008) as "real estate fever", in the last few decades, the place underwent the first external interventions through the installation of secondary residences. In the meantime, deforestation has increased on the slopes of the Serra Geral de Guaratiba and also in some isolated areas of the lowlands, in order to make way for the farms that housed the houses used for vacations and weekends. Many of the streams that flowed down the massif, bringing crystal-clear water to the flat areas, were dammed upstream to form lakes and natural pools on the new estates. These rivers, which until then had been populated by fish - such as moths, caraûnas, piabas etc. - and crustaceans such as lagostas and pitus (large freshwater shrimp), are gradually losing their fishiness.

Subsequently, the increased influx led to an even greater impact, as many of the old streams began to receive untreated sewage, turning them into veritable black ditches. As this process unfolded, many hills were demolished, contributing to land and real estate speculation, giving way to new homes and serving as landfill for new subdivisions in an area that until then had not suffered a major impact: the tidal plain.

It was from the 1990s onwards that environmental degradation took off. The constant landfilling of mangrove areas and alluvial channels, with the aim of building condominiums, caused unprecedented impacts on the mangrove biosystem, resulting in a sharp decline in the population of crabs and guayamus, crustaceans that represent a source of income for many families of pickers.

Due to the lack of planning and infrastructure, the changes discussed here do not in fact represent a new urban trend, but rather a radical change, which Guaratibanos point to as one of the main factors responsible for the degradation of the place's natural setting. This phenomenon, however, by changing the spatiality in question, can also change people's lives, since the influences between people and their places are reciprocal.

The growth of cities and the urbanization of the world is, according to Wirth (1976), one of the most remarkable facts of modern times. The social scientist recalls that the change from a rural society to a predominantly urban one took place in the space of a single generation in some central countries. The author, however, is also concerned with the difference between urbanism and urbanization, where urbanization refers to the urban form, represented by buildings, and urbanism, to the peculiar lifestyle of cities. Many other theorists have also addressed the issue of the distinction between the city and the urban, since common sense usually refers to these terms as synonyms.

Because it represents a hyper-complex reality and has several dimensions, the urban is represented in different ways. In this sense, Yàzigi (2003), in his discourse on the various values that surround the urban environment (historical value, social value, economic value, affective value, etc.), stresses the importance of a sense of belonging in building a permanent urban heritage. According to Yàzigi, without a sense of belonging you can't expect much from an urban agglomeration, which is increasingly becoming a modern form of camping. Affective value therefore represents an indispensable condition for building an urban environment that is understood as heritage and as a topophilic environment (TUAN, 2012).

For Corrêa (2000), the need for greater consumption of space as a result of land appreciation, which is responsible for the rise in value of real estate in the central areas, points to much more advantageous conditions in the distant peripheries, which are endowed with amenities. Under these circumstances, the urbanization of a peripheral part of the city of Rio de Janeiro like Ilha de Guaratiba, much more than a mere spatial change, represents subjective transformations in which its residents come to experience a different lifestyle, based on new values. In this sense:

urbanization no longer denotes merely the process by which people are drawn to a place called a city and incorporated into its system of life. It also refers to that cumulative accentuation of the characteristics that distinguish the way of life associated with the growth of cities and, finally, to the changes in meaning of the ways of life recognized as urban (WIRTH, 1976, p. 93).

For Tuan (2012), the suburb is an ideal, as it suggests a perfect lifestyle in which the best of rural and urban life are combined without their defects. In pursuit of this alternative lifestyle, more and more people are moving to Ilha de Guaratiba. This mobility, however, is set to increase even more with the construction of the "Grota Funda Tunnel". Let's consider José Maria's opinion on this matter:

The tunnel is one of the bulwarks of the change we see in the place, there's no doubt about that. Its construction facilitated access to Recreio and Barra, since the mountains had always been a natural obstacle. The Grota Funda tunnel, by facilitating access to the area, contributed to an even greater densification of people and buildings. In addition, its drilling served to introduce Guaratiba Island to Rio de Janeiro. However, when the tunnel introduced the island to Rio, cariocas - who didn't know it - discovered that this is a very good place to live. For this reason, **the Grota Funda tunnel** represents the beginning of the end of this **"paradise", in** other words, **the gateway to hell** (José Maria - rural producer - 70 years old).

In the course of the process of expanding Rio's urban fabric, the drilling of the city's tunnels often represented the incorporation of new areas into the metropolitan context (CARVALHO, 2004). In this context, the testimony of resident José Maria is emblematic, since it alludes to a new reality, namely the connection of Ilha de Guaratiba to the urban context of Rio de Janeiro. The place, however, is still in transition. In the midst of its living niche, there are persistent rural characteristics of yesteryear in the midst of modern urban trends.

Although he comes close to phenomenology and hermeneutics in some of his research and conferences, the geographer Joâo Rua does not share this humanistic approach. However, in order to elucidate the issue related to the multiple interactions between rural and urban areas, we believe his assertion is pertinent. In his research, this thinker (RUA, 2002 a; 2002 b) emphasizes both the question of the manifestations of the urban in the rural, a phenomenon he calls urbanities, and the strength of the rural in the face of the urban (ruralities). For Rua, the new ruralities command the current process of spatial restructuring, since the rural, when incorporated by the phenomenon of urbanization, is integrated into the urban while retaining some specificities.

In his theoretical considerations about urbanities and new ruralities, Rua (2002 A) moves towards approaches that refer to the hybridity of these new spatialities (SANTOS, 2002), since the two realities (con)merge. Thus, the "old" or "internal", represented by the past rural structure of Ilha de Guaratiba, when associated with the "new" or "external", portrayed by its modern urbanization process (SANTOS, 1997), give the place a range of new characteristics, specificities, particularities, singularities, as well as a broad niche of recent and past symbols that present themselves to our interpretation.

5 AN ISLAND OF SYMBOLS

A symbol has the power to suggest a whole, it transcends its condition as such and, as an integral part, it blends in with the place in which it is located. In this respect, the symbolic charge of a temple or stadium can be much broader, more expressive than its original purpose. In fact, the cross symbolizes Christianity, the crown the monarchy (TUAN, 2012), just as the Brandenburg Gate represents one of the greatest symbols of the German nation (FREITAS, 1999). Symbolism, understood as an emblem or interpretation of the meaning of a particular symbolic element (symbol), has emerged in recent decades as an extremely important concept for humanistic and cultural research - studies related to understanding the subjective dimension of place (TUAN, 1980; 2012; MELLO, 2000, 2003; FERNANDES, 2015). According to Cosgrove (2004):

> All landscapes have symbolic meanings because they are the product of man's appropriation and transformation of the environment. Symbolism is most easily grasped in the most elaborate landscapes - the city, the park and the garden - and through the representation of the landscape in painting, poetry and other arts. But it can be read in rural landscapes and even in the more apparently non-humanized landscapes of the natural environment. The latter are often powerful symbols in themselves

(COSGROVE, 2004, p.108).

To understand the expressions imprinted by a culture on its landscape, we need knowledge of the language used: the symbols and their meaning in that culture. For the author, despite the very tenuous link between the symbol and what it represents, all landscapes are symbolic. By emphasizing that human scenarios are loaded with symbolism, Cosgrove focuses on nature and the natural landscape as powerful symbols in themselves, based on the assumption that any human intervention in nature involves its transformation into culture. Although this transformation is not always visible, especially to an outsider, the natural object becomes a cultural object when it is given a symbolic meaning (COSGROVE, 2004). Let's now look at Tuan's (2012) words:

> Of the many and varied reasons for moving to the suburbs, the search for a healthy environment and an informal lifestyle are among the oldest. We have repeatedly observed how the feeling for nature and rural life is encouraged by the pressures of urban life. The city environment is both seductive and irritating, beautiful and unpleasant. The rich have always been able to escape this by going off to rest in their country houses. In the Western world, the feeling for nature culminated in the romantic movement of the 18th and 19th centuries (...). The city symbolized corruption (...). The countryside symbolized life: life revealed in the fruits of the earth, in the green things that grow, in pure water and clean air, in the healthy human family (TUAN, 2012, p. 324-325).

Symbolism is not restricted to centers of wellbeing, affection, deprivation or experience, as vast, strange and distant spaces are configured as symbols of rejection (TUAN, 1980; 2012; MELLO, 2003). Understanding symbolism as the mark of an idea - both negative and positive - of a certain symbolic element, Tuan (2012) proposes a counterpoint between the city and the countryside, suggesting that since the industrial revolution, the city - little by little - has ceased to symbolize an ideal of life, giving way to the countryside through a return to a feeling for nature. According to Tuan, by acquiring some of the values of the countryside, the suburb - understood as the frontier of metropolitan expansion - comes to represent an ideal, as it suggests a perfect lifestyle, in which the best of rural and urban life are combined without their defects (TUAN, 2012). In this sense, whether in the American context above or in the Brazilian suburbs, especially in Rio de Janeiro, the metropolitan peripheries come to represent a symbol of well-being for their residents (CORRÊA, 2000; FERNANDES, 2006; SOUZA, 2005).

Places and symbols acquire deep meaning through the emotional bonds woven over the years. The place itself is a symbol of affection, well-wishing, satisfaction, happiness and togetherness, on the one hand, but also the scene of struggles and day-to-day life. The symbolic nature of places establishes connections by decoding and translating their past and connecting it to the present (MELLO, 1990; 2003).

Still on the subject of the symbolic universe, let's consider the findings of geographer Doreen Massey (2008):

> And so there is "place". In the context of a world that is certainly increasingly interconnected, the notion of place (usually referred to as "local place") has acquired a totemic resonance. Its symbolic value is constantly mobilized in political arguments. For some, it is the sphere of everyday life, of real and valued practices, the geographical source of meaning, vital as a fulcrum while "the global" weaves its ever more powerful and alienating webs. For others, "a refuge in place" represents the protection of drawbridges and the construction of walls against new invasions. Place, in this reading, is the place of negation.

In an attempt to translate the symbolic value of place, Doreen Massey (2008, p.24-25) discusses its wide range of meanings. From her perspective, place symbolizes - among other things - the sphere of everyday life, the geographical source of meaning, a vital point of support, as well as representing refuge and protection from the powerful and alienating webs of the global. Advocating a new stimulation of spatiality, the author points to nature and the natural landscape as symbolic foundations for recognizing place (MASSEY, 2008).

Delving into this universe of meanings and values, Joel Bonnemaison (2002, p.109-111) emphasizes:

> A geosymbol can be defined as a place, an itinerary, a stretch of land that, for religious, political or cultural reasons, in the eyes of certain people and ethnic groups takes on a symbolic dimension that strengthens their identity (...). Symbols gain greater strength and prominence when they are embodied in places. Cultural space is a geosymbolic space, charged with affectivity and meaning.

The geosymbol - a concept developed by Joel Bonnemaison (2002, p.109-111) - can be understood as a place-symbol, loaded with affectivity and meaning. Among the premises defended by geographers of the humanistic horizon are those related to the symbolic content of places (COSTA, 2008) and the mosaic of symbols that reside there (MELLO, 2003, 2008). In this sense:

> The symbolic nature of places reveals itself to human beings as something that precedes language and

discursive reason, thus presenting certain aspects of reality, emphasizing the relationship between the symbolic and the place. These relationships are mediated by symbols, which can be a material reality that is linked to an idea, a value or a feeling. We therefore understand that symbolic mediations permeate personal attitudes towards places (COSTA, 2008, p 149).

In his discourse on the issue of cultural heritage as a set of symbols that refer to the memory of the place, Costa (2008) alludes that "the symbolism of places leads us to the concept of vernacular landscape where this character is made explicit in the set of representations, both of ancient landscapes and current ones, expressed through the knowledge and actions of man" (p.151). For the author, certain natural or cultural elements, when associated with the daily relationships of individuals or social groups, can define a set of symbols that express the memory of the place. In these circumstances, everyday relationships and the consequent understanding of places and their symbols can turn a space into a place, once it has been affected. "In this context, place becomes more interesting as a reference for identity and at the same time acquires a symbolic value" (COSTA, 2008, p. 155).

Considering the symbolic universe of places, reconciling, understanding and decoding the past and present symbols of Ilha de Guaratiba are tasks to be undertaken in the following pages, starting with the symbolic value of its toponym.

5.1 **Deciphering toponymy in its meanings**

Naming places has a strong meaning, since these names establish connections between the place in the past and today. As such, toponymy reveals ownership, memory, symbolism, affection, adherence, resistance and intimacy with the place named (MELLO, 2007). In this field, Corrêa (2003, p. 176) emphasizes: "toponymy is a relevant cultural mark and expresses the effective appropriation of space by a given group, being a powerful element of identity". Under these conditions, the names of streets or neighborhoods give places a strong identity, being the result of experiences, clashes, utopias and values, in the midst of friends, relatives, strangers, acquaintances and feelings, making up a whole of introjections, strangeness, adherence and belonging (MELLO, 2000).

For Lessa (2001), toponymy is the first and most faithful record of places, a kind of baptism. According to the author, people's loyalty to their names takes on a powerful dimension. In this line of argument, Lessa (2001, p. 58) argues:

> Names have a much longer lifespan than the material configurations of places. One example is the names that endure even though their original places no longer exist: they retain an unmistakable symbolic character. Praça XI, which is still evoked in Rio's samba; Castelo is the site of a hill that has already been demolished; Rua do Ouvidor, which almost nobody knows who it was, and there are even doubts as to which ouvidor it might have been. Other places change euphoniously: the Battle of Cerro-Corà, from the Paraguayan War, gave rise to the Serra Coral slum. The Praça do Asseca became Praça Seca; the sitio do Willian became Ilha (de Guaratiba) etc.

Showing himself to be a profound connoisseur of the places in his city and the contexts that gave rise to their names, Lessa points in the direction we are focusing on, since he approaches toponyms as a true mosaic that unites contemporary and past elements. In his quest to understand the meaning and significance of the names of different places in the city of Rio de Janeiro, the author discovered that it was "a certain Willian, who owned a place in Guaratiba, who was responsible for the origin of the name of the place called 'Ilha'" (LESSA, 2001, p. 427).

Like most people who look at Ilha de Guaratiba, Lessa used the writings of historian Rivadavia Pinto (1986) to arrive at the origins of the place's name. According to Pinto, the place name "Ilha" originated as a corruption of the name of the Englishman Willian. Coming as part of the English escort that protected the Portuguese Royal Family on their transfer to the former Terra de Santa Cruz in 1808, William took possession of and began to live in the area that is the subject of this investigation. As the natives were not very good at pronouncing his name correctly, they began to call him "Wilha", his "Ilha de Guaratiba" and, finally, "Ilha de Guaratiba", the former owner of the local domains (LESSA, 2001). The much older toponym "Guaratiba" was derived from the large number of wading birds that inhabited the area - the guaràs. As the word "tiba", in Tupi-Guarani, means abundance, Guaratiba, etymologically, means "abundance of guaràs" (PINTO, 1986). In these terms, the toponym Ilha de Guaratiba arises from the crossing of British and indigenous words.

Although this corrupted version is widely accepted, there is another version that tries to elucidate why a place not surrounded by water is called an island. Let's see:

> On the old maps of Ilha de Guaratiba, it was possible to see the two rivers that skirted the area, turning the place into an island as it was surrounded by their waters. At that time, not only were the waters of our rivers more voluminous, but they were also much more influenced by the tides. The story of "Seu Ilha" is the most credible. However, there are other stories about the place's name. The old Guaratibanos used to say that in the past this was an island surrounded by tidal channels. Many still say today that these rivers were very large. This large volume of water increased even more when the tide rose and the people here on the island were stranded during periods of high tide. The old-timers say that the existing rivers and canals made it difficult for people to reach the town. At that time, the place became isolated from time to time, depending on the flow of the rivers and tidal channels. According to this version, the place was periodically "cut off" by the floods caused by the tidal channels. There was even a small port here where the agricultural products produced in the mills were transported by way of the portinho river (Pipo - biologist - 49 years old).

Everyone often agrees that Ilha de Guaratiba is named after an Englishman. Resident Pipo, however, looks to the roots of the place to find another explanation for the name. Although he doesn't dispute the official toponymic version based on William's (Seu Ilha) story, it's worth noting that his previous account presents a different idea about the place's name. In his testimony, Pipo points out that the place name "Ilha" is linked to the periodic flooding of part of the Baixada de Guaratiba, due to the rise of the water at high tide. The resident bases his theory on historical records that confirm the occurrence of regular flooding in the areas close to the canal rivers of the Guaratiba Tidal Flat (CASTRO, 2002; PINTO, 1986) and on archaeological (KNEIP, 1987) and biological (ARAÙJO, 1987; MENEZES, 2005) studies of the Baixada de Guaratiba. Pipo also mentions the occurrence of an old port on Ilha de Guaratiba, which would prove the existence of flowing rivers in the area. The biologist's information is echoed by Fridman (1999), who refers to the existence of the Ilha Port, which was in full operation at the beginning of the 19th century.

Let's consider the thoughts of resident Rogério - chemist - 44 years old. When asked about the name of the place, Rogério emphasizes:

> There is a project to change the name of our place to Recreio de Guaratiba. I think this idea is absurd. Our name is a fundamental part of our identity. The name Ilha de Guaratiba reflects what the place is: an island surrounded by mountains, greenery and tranquillity.

The power and importance of the place name "Ilha" can be detected in the above-mentioned account. In it, Rogério shows his indignation at Decree No. 24230 of May 20, 2004 (Annex 5), a project by then mayor César Maia that intended to turn Ilha de Guaratiba into an "extension of Barra" (FERNANDES, 2006). The former mayor's aim was to turn the area into a planned residential neighborhood, to be called "Recreio de Guaratiba" (figure 4). A strong identification with the place and its name is evident when Rogério repudiates the replacement of the toponymy, in the case of Ilha de Guaratiba, revealing memory and identity.

Figure 4 - Recreio de Guaratiba neighborhood, according to Decree No. 24230 of May 20, 2004 - Mayor César Maia's administration.

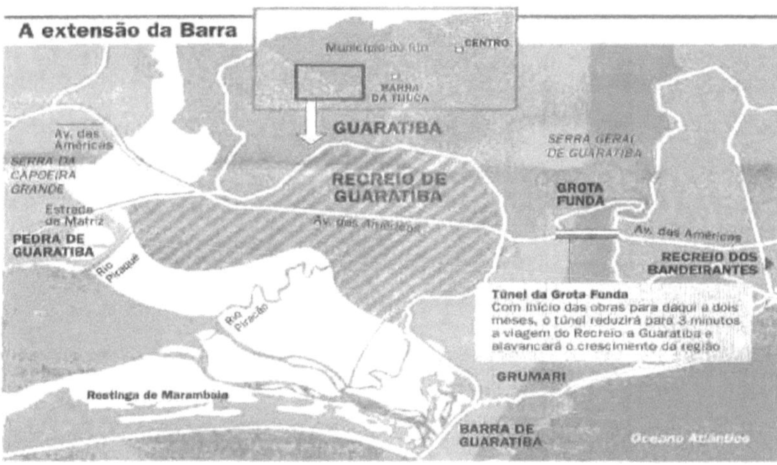

Source: PORTAL Guaratiba. Available at: <http://www.portalguaratiba.com.br>. Accessed on: November 20, 2015.

Because it isn't officially considered a neighborhood, Ilha de Guaratba - the neighborhood chosen by its residents - is often ignored, and is thus included in Guaratiba in general, as if it didn't have its own identity. However, the insiders, explorers and deep connoisseurs of their lived universe feel offended when their place is coupled with another that, due to rejection/hatred, in the war between places, can be confused with a space they don't identify with.

The following are the words of a teacher from Ilha de Guaratiba:

Honestly, I can't understand why Ilha de Guaratiba isn't considered a neighborhood by the City Hall. As well as having its own identity, linked to our agricultural past and our natural attributes, our boundaries are well demarcated, both by human occupation and by the Serra da Grota Funda. In my opinion, the characteristics that distinguish us from Guaratiba as a whole and from our neighboring districts are clear (Sandra - teacher - 57 years old).

In her fervent adherence to the place, Professor Sandra shows her indignation at the fact that Ilha de Guaratiba does not have autonomy as a neighborhood from the local authorities. As you can see, for her the place deserves to be elevated to this category because of its distinct physical and cultural aspects.

"Limits, borders and obstacles are issues that are frequently focused on in geographical studies" (MELLO, 2001, p. 89). Even informal geographers (Lowenthal, 1982), like Sandra, mention these themes when they want to demonstrate the location of their lived universe. Let's repeat part of her speech: "How come the neighbourhood hasn't been officially legitimized yet? Why hasn't it been dismembered from Guaratiba?" For her, Ilha de Guaratiba "has a life of its own" with boundaries and identity - essential, fundamental elements even in the defense of her lived place.

In the words of Mello (1991, p. 151), "walls of different kinds separate" or segregate the individual in their world. Some are natural, like mountains. Others are created by culture or forged by the human mind. "Within the walls, man, in his private and/or collective world, united with his own" and bound by topophilic ties to the surrounding environment, "is invited to carry out his daily activities". Beyond its limits, the world is frightening, uncomfortable, far from the fraternity that can be felt in the lived universe. Mello's considerations (1991, p. 151) can be seen in the repertoire belonging to orality, in testimonies such as the interview with teacher Sandra, who emphasizes Ilha de Guaratiba's "own identity" and the "well demarcated limits, both by human occupation and by the Serra da Grota Funda".

In fact, exploring Professor Sandra's universe of ideas, it's worth noting that Guaratiba is represented by the administrative region of the same name, and is made up of three neighborhoods: Pedra de Guaratiba, Barra de Guaratiba and Guaratiba, the latter being the largest neighborhood in the municipality of Rio de Janeiro, and in which Ilha de Guaratiba is located. Despite the vastness of this area, qualitative research has shown that it is on Ilha de Guaratiba that the elements mentioned here, relating to local toponymy, are strongly characterized, giving the locality a solid identity (figure 5).

Figure 5 - Guaratiba Island - location.

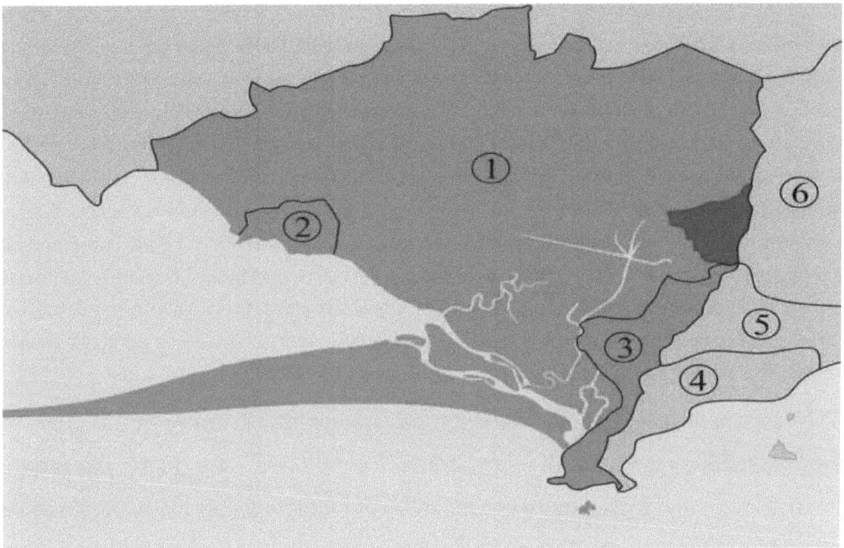

Caption:

→ Guaratiba Island

①→ Guaratiba

②→ Pedra de Guaratiba

③→ Barra de Guaratiba

④→ Grumari

⑤→ Recreio dos Bandeirantes

⑥→ Vargem Grande

Source: FERNANDES, 2003.

The guarà, another example, is not evoked as a symbol in Guaratiba as a whole. But on Ilha de Guaratiba, the bird that gave the place its name symbolizes a return to the past and a veneration of nature.

5.2 **The symbolism of natural elements**

Let's look at another discourse concerning the symbolic content of Ilha de Guaratiba:

Our natural elements have a great deal of symbolism. Here on Ilha de Guaratiba, when we look around us, we come across greenery and mountains that provide us with an exuberant landscape. As one of the symbols of the place, I would choose the Serra da Grota Funda. I think the intertwining of green and mountain is perfect. I stand in my backyard enjoying that beautiful green mountain. For me, this is our greatest symbol. The guarà is another symbol of the place. The bird represents our exuberant nature. When we talk about the symbols of a place like Ilha de Guaratiba, we can't help but consider the origin of its name. The name Guaratiba comes from the many Guarâs who inhabited the area. Scientific birdlife records show that until the 1980s there was one remaining pair of guarâs in the area. These birds were killed by a researcher who needed to record the occurrence of the species in the area. This generated a great deal of controversy, but scientifically proved that the guarâs originated in Guaratiba. The fossils of these slaughtered birds are even on display in the national museum as the last guarâs from Guaratiba. I consider the guar to be the symbol of Ilha de Guaratiba because it is linked to the name of the place and associated with its nature (Marcelo - university lecturer - 36 years old).

The statement above alludes to the most emblematic symbols of the place in the understanding of one of its residents. For Marcelo - the guarà, the green, the mountain and the scenery produced by the interweaving of the most visible elements of the natural landscape are the marks that best represent the place (FIGURE 6).

Despite today's notorious veneration of the bucolic environment, as noted in the words of the resident interviewed in previous surveys (FERNANDES, 2003; 2006), what today represents an ostentatious symbol was an expression of displeasure, since it consigned the place to a condition of backwardness in relation to the urbanized districts of the city. In these terms, the rural and country atmosphere of Ilha de Guaratiba, especially for young people, consigned the neighborhood to a pretentious stage out of step with the rest of the city. In an attempt to elucidate the reasons why a particular artifact or place adds or decreases value to its characteristics, Mello (2003) advocates that a symbol loses or receives this distinction depending on the darkness or clarity attributed to it over time. Therefore, as society and culture change over time, the relationship with the environment can also change, even moving towards a true veneration of nature (TUAN, 2012).

Figure 6 - Serra da Grota Funda or Serra Geral de Guaratiba, an elevation belonging to the "Pedra Branca Massif" that separates Ilha de Guaratiba - to the west - from Recreio dos Bandeirantes.

Source: REDONDO, 2012

The first human evocations in relation to nature remind us of fear and aversion to a hostile environment, where anthropic vulnerability was evident in the face of a wild habitat in which man was notoriously unfit to live (PARK, 1976; TUAN, 2005). The first cities and/or human settlements, such as Jericho (in Palestine) and Ur in ancient Mesopotamia (now Iraq), arose millennia before Christ and were protected from enemy armies and the many dangers of nature by great walls (SOUZA, 2005). It took the emergence of the great cities at the time of Alexander the Great for there to be a strong reaction in favor of the rusticity of natural environments. In these steps, when a society reached a certain level of development and complexity, people began to observe and appreciate the relative simplicity of nature. This kind of feeling only emerged after the construction of cities, when the pressures of urban life made rural peace and a romantic appreciation of nature attractive (TUAN, 2012). Nature, once a symbol of a topophobic environment (TUAN, 2005), today represents an indispensable attribute for the enhancement of areas within the urban environment and an essential merit for the pleasantness of places. In this respect, the tripod of amenities, or just one of its items - sea-green-mountain - nowadays constitutes an element of value in large cities (ABREU, 2008; ASSIS, 2003; CORRÊA, 2000; MELLO, 2007).

The scenery of ennobled areas, boasted by the wealth of green sea-mountain amenities, today represents a symbol of living well and status, in general. In Ilha de Guaratiba, which is endowed with mountain greenery, these predicates are part of the memory of those who are rooted in the place and, more recently, attract and adjust to the expectations of the new residents.

Like any other part of space, Ilha de Guaratiba has its own specific characteristics, which are responsible for distinguishing it from other places. In this context, its natural attributes are a clear example of its particularity. In fact, being a lowland area with ample and unique greenery, surrounded in large part by the Pedra Branca Massif (Serra Geral de Guaratiba), the place has as its hallmark its exuberant natural landscape, provided by the green-mountain junction (FERNANDES, 2003, 2006, 2009).

According to Guerra (1993), relief is one of the fundamental elements of the physical area and its shapes and layers are studied by geographers. Because it has mountains and plains in its domain, the Ilha de Guaratiba neighborhood has been the focus of research by these specialists. Over time, the local morphology has exerted a great influence, transforming Ilha de Guaratiba into a place in its own right. Its topography is basically composed of a plateau area (Planicie da Maré de Guaratiba) - where some "barriers" appear in the form of isolated rounded reliefs (FERREIRA and OLIVEIRA, 1987). The local morphology blends in with the vegetation of the lowlands, slopes and massifs, thus providing a graceful view, both from the top of the massif, from where we can glimpse a rich panorama, and from the plateau itself, where when we look at the mountain range, we are faced with the picture presented by the hills that surround the place.

In terms of greenery, Ilha de Guaratiba is privileged. It is home to important environmental preservation areas with significant stretches of Atlantic forest. The green predominating on the plateau, combined with the forested hills, gives the place a landscape of "indescribable beauty", as Magalhâes Corrêa pointed out in his wanderings through the Rio de Janeiro back then (CORRÊA, 1936, P.192). The aforementioned researcher's description, which dates back to the 1930s, is still referenced today by Guaratibanos who also mention these elements and the landscape they provide as the most important symbols of their lived place.

For decades, natural domains have been valued by real estate agents and used as a kind of lure by property speculators (ABREU, 2008; ASSIS, 2003; CORRÊA, 2000; MELLO; 2007). In addition, the bucolic landscape holds a significant symbolic charge for individuals and social groups, and in the meantime has a strong influence on both the construction of identity and the formation of place (COSTA, 2008; MASSEY, 2008).

In his previously mentioned account, university professor Marcelo emphasizes his relationship with nature (the green and the mountains), making it clear that the natural setting of the place exerts a certain fascination on him. For his part, even the mention of the guarà as a symbol of the place reveres the exuberant nature of Ilha de Guaratiba. Phrases such as "this is our greatest symbol" denote a clear sense of belonging, where, in the midst of introjections, it is not possible to separate the individual (Marcelo) from his place-symbol (Ilha de Guaratiba).

As we continue to unveil the symbolic universe of Ilha de Guaratiba, let's consider the opinion of another of its residents:

> It may sound silly, but in my opinion, the elements that best represent our place are the crab and the heron. Here, mangroves flourish all over the lowlands. The quantity of crabs and herons is remarkable in this environment and throughout the locality. Another apparently simple thing that catches my eye is the fact that the island is the only place where you can still find "real bananas", grown naturally on its hillsides and without the addition of carbide in the ripening process. Because of its huge banana plantations and the quality of the fruit produced, I choose the silver banana as one of the symbols of Ilha de Guaratiba (José Maria - rural producer - 80 years old).

Because they are intimate, private and individual, symbols can vary from person to person, and unusual symbols, such as the ones above, can emerge through personal experiences. José Maria lived in Recreio and got to know Ilha de Guaratiba through the gardens that sell ornamental plants. In 1993, he bought a property and started living there permanently. Living in his new environment allowed the farmer to identify with one of the elements belonging to the niche characteristics of the place: the silver banana - grown on the hillsides.

Over time, the silver banana took the place of coffee on the slopes of the Serra Geral de Guaratiba after the decline of coffee growing. Until the 1980s, it was common for people to go up to the hills of Ilha de Guaratiba, where the silver banana could be picked ripe without restriction, due to its abundance. At that time, the island produced a large quantity of this product, which was then sold at street markets and CEASA. Nowadays, although it no longer produces as much as it used to, the silver banana is still one of the symbols of the town and is sold (in natura) on the roadsides.

José Maria also cites crabs and herons, representative species of the mangrove biosystem, as symbols of the place. The crabs live hidden in deep burrows in the areas close to the coast. Even after part of the Guaratiba lowlands have been landfilled to make way for condominiums, it is still possible to find a large number of crabs in the local mangroves. Similarly, herons are quite common in the area. Many condominiums in the area allude to this bird in their names. Examples are Condominio Recanto das Garças and Residencial Reserva das Garças.

Because of its charms, Ilha de Guaratiba can be considered a great agora, where the elements of nature have been chosen as the ones most responsible for giving the place its main specificities. The aforementioned physical and landscape characteristics of the place have also become its most visible symbols and, for this reason, the most evoked. However, there are still symbols in the space in question that take us back to different times. As is common in places that undergo spatial changes, in Ilha de Guaratiba we find elements that represent the past and others that take us back to the present. With regard to the symbolism of yesteryear, it is important to note that, even though they no longer represent a visible and/or pulsating reality, these symbolic elements still permeate the nostalgic minds of Guaratibanos. With regard to today's symbols, the range of meanings that manifest themselves in the lives and relationships of the residents of Ilha de Guaratiba is striking.

5.3 Unveiling past and present symbols of Ilha de Guaratiba

A common human tendency is to anchor the range of attachments, belongings and feelings in our place to certain experiences and symbols from the past (MELLO, 1991; 2000; 2003; 2008; TUAN, 1982; 2012; 2013; BUTTIMER, 2015). In this vein, let's consider the following words:

> I miss the rural life that has been disappearing as the place has been invaded by unbridled urbanization. I miss the orange groves. When I was a child, it was a great adventure to "steal" oranges from Mr. Valdir's plantation. From my house I could see the orange trees full and I couldn't resist. The games played in those days: rounders, flag football, burnout, among others, brought people closer together. Today, digital games and the internet are the norm, increasing individualism and distance between children and people in general. A rural atmosphere enveloped us. These characteristics have always marked our place and are being lost over time. The closeness between people was much greater. There were few families and we knew everyone by name. I miss that time (Dalva - housewife - 68 years old).

In her testimony full of nostalgia and pride, resident Dalva recounts some of the events, experiences, games, characteristics and past symbols that remain alive, pulsating and ever-present in her memories. The immaterial symbols of Ilha de Guaratiba are exalted because of the beauty of the place's country life in memorable times. Because they pulsate vigorously in the memory of nostalgic people who experienced them in their moments of splendor, they are recounted with a clear sense of regret. Dona Dalva would certainly like these symbolic characteristics of the place not to be lost in the vortex of time. Her testimony, shrouded in melancholy expression, refers to cordiality, bucolicism, as well as interaction, friendship and other characteristics directly linked to interpersonal relationships. This in itself demonstrates the nature of this range of symbolic immaterial goods linked to and introjected into the past conjuncture of its lived place.

Many thinkers share the idea that the spatial and social changes that have taken place in most communities lead their residents to think that "lost ways of life are the right ways to live" (GONDAR and DODEBEI; 2005, p. 9). An individual almost "never values what they have until they lose it" (BUTTIMER, 2015, p. 5). This nostalgic feeling is usually associated with older people (NORA, 1993; BOSI, 2003). This is the case of Dona Dalva, who, with 68 years of experience on Ilha de Guaratiba, has many memories of the place in different temporalities. Her memories reveal geographies of yesteryear which, although they are related to past spatialities, remain alive in her memory and in her heart. They are "memorable geographies" that he misses.

Both places and symbols can be forged through immediate experiences. At first glance, an individual may fall in love with a particular spot. On the other hand, they may need a certain amount of time for an artifact to capture their attention, rising to the level of symbol or place (MELLO, 1991, 2003). In highlighting place as a reservoir of memories and dreams, Tuan (2013, p. 227) stresses: "the sense of time affects the sense of place". Let's look at an excerpt that points to the relevance of the past as a pillar of individual and collective identity and a source of meaning and symbolism:

> What can the past mean for us? People look back for various reasons, but one is common to all: the need to acquire a sense of self and identity. I am more than what is defined by the fleeting present. I am more than someone who at this moment struggles to put thought into words: I am also a writer whose book has been published, and here the book is, bound, by my side, renewing my confidence (...). To strengthen our sense of self, the past needs to be rescued and made accessible (TUAN, 2013, p.227228).

Each person carries their place with them through experiences, familiarity, affection, belonging and other experiences. This range of feelings is woven over time and evoked, consciously or unconsciously, at all times, denoting that what we are and what we possess is the result of our history and the mosaic of experiences we have lived through in our home, understood on the most diverse scales (BUTTIMER, 2015). In this sense, "awareness of the past is an important element in the love of place" (TUAN, 2012, p.144) and history is responsible for the feeling of belonging and love for the sphere of the lived, since, over time, a person invests a significant part of their emotional life in their home and neighborhood (TUAN, 2012).

Human beings try to understand the world based on their past experience (LOWENTHAL, 1982; 1998), assimilating elements presented by formal and informal education over the course of their lives. This is, as philosophers argue, the so-called stock of knowledge (SCHUTZ, 1979). In these circumstances, "experiences in the scenarios of the past are treasures kept with great tenderness" (MELLO, 1991, p.235). In line with this premise, Harvey (1992), drawing on Rossi, cites the historical reference and the collection of the past as a source of meaning for "cultural symbols":

> The impulse to preserve the past is part of the impulse to preserve the self. Without knowing where we've been, it's hard to know where we're going. The past is the foundation of individual and collective identity; objects from the past are the source of meaning as cultural symbols. The continuity between past and present creates a sense of sequence to the random chaos and, as change is inevitable, a stable system of organized meanings allows us to deal with innovation and decay. The nostalgic impulse is an important agent of adjustment to crisis, its social emollient, reinforcing identity when confidence is weakened or threatened (HARVEY, 1992, p. 85).

Since every lived experience dates back to the past, the importance of the experiences, places and symbols of yesteryear in the process of building identity is unquestionable (HARVEY, 1992). This heritage links people to their lived place, which is now proclaimed and revered not only for its contemporary characteristics, but also for the history and geography that have been built by individuals and social groups on the ground they have experienced over time. As already explained, "history plays an essential role in the human sense of territoriality and place" (TUAN, 1982, p. 156). At this point, "the identity of a place is its physical character, its history and how people make use of their past to promote awareness" (TUAN, 1982, p. 156).

Returning to the symbolic elements of Ilha de Guaratiba, expressed in Dona Dalva's testimony, proximity and contact between neighbors are the basis for the simplest and most elementary form of association we deal with in organizing community life. However, under the complex influences of urban life, what can be called the feeling of neighborhood has undergone radical changes, producing different types of local communities. In this sense, we can say that there are nascent neighborhoods - in a phase of adaptation - and neighborhoods in a stage of dissolution and metamorphosis. In city environments, as Park (1976) and Simmel (1976) point out, neighborhood ties tend to lose a lot of importance. In contrast, in simpler communities there is an intense life of relationships. Thus, the reserve, indifference and blasé air manifested in human relationships can be understood as instruments used by individuals to immunize themselves against personal demands and the expectations of others (WIRTH,

1976).

Superficiality, anonymity and the transitory nature of urban-social relations also explain the sophistication and even snobbery generally attributed to city dwellers (PARK, 1976; SIMMEL, 1976; WIRTH, 1976), especially in noble neighborhoods, with higher income strata, downgraded in the midst of an ethnocentric disdain which, in a way, was already being signaled in Ilha de Guaratiba, due to the growth in income and self-segregation.

With regard to the aforementioned characteristics of the urban way of life, linked to the "disappearance" of the feeling of neighborhood and the corrosion of the traditional basis of social solidarity, Corrêa (1992, p. 34-35) argues:

> Densification, because it implied significant inter-neighborhood residential mobility, resulted in the disappearance of neighborhood units where everyone knew each other and, when necessary, helped each other (...). Associated with this change is the growing anonymity of the population living in middle-class neighborhoods. Anonymity that includes a certain amount of distrust and fear of the other.

As we can see, for decades, researchers have been looking at issues related to the differences that occur in relationships. In central places and/or urban localities, these relationships are generally formal and can be cold and indifferent. In peripheral neighborhoods and/or community places, on the other hand, personal relationships denote closeness and a strong sense of neighborhood, as evidenced by the opening of doors and the streets or alleys experienced as an extension of the home. As far as Ilha de Guaratiba is concerned, a place in transition, the close ties of neighborliness are gradually being replaced by estrangement between people. This evidence is reported by experienced residents such as Dona Dalva. Dalva's testimony offers us support to understand an increasingly common situation in the midst of the inexorable march of urbanization that we are witnessing: the estrangement and non-recognition of old residents in relation to new residents. This phenomenon is becoming increasingly common in places such as Ilha de Guaratiba, located on the peri-urban boundary. Although Dalva's words are linked to the changes taking place in a metropolis, they have an explicit connotative content. By interpreting the story in question, we can understand that spatial transformations can lead to existential metamorphoses. Assertions like this, however, only become possible through qualitative research. Through interviews or informal conversations, it becomes possible to explore people's private worlds. In this way, different symbolic elements of Ilha de Guaratiba can be explained by considering the adherences and other feelings of belonging that, in the midst of introjections, address even the individual to an idyllic past.

In this niche of appreciations, trying to decode the thoughts of the resident Dalva, when a place reaches a certain level of development and complexity, people tend to give more value to the relative simplicity of the old habits. These, however, are overtaken by new relationships - no longer based on camaraderie, exchange and solidarity - but on distrust, fear and distance between people (TUAN, 2005; 2012). When members of a particular social group or community realize that changes are occurring too quickly, the longing for an idyllic past increases significantly (TUAN, 2013).

Since "meanings emerge from the deepest experiences that accumulate over time", this repository of meanings becomes a symbol (TUAN, 2012, p.203). In this shortcut, the role of the neighborhood, which reveals the bucolic nature of the recent past, is evoked by Guaratibanos like Dona Dalva, revealing symbolic characteristics of the place. Although it has lost the depth of other times, proximity - which forged identity, feelings, familiarity and reciprocity - remains present in the hearts and nostalgic minds of those who were linked to the place through their experiences there.

Time passes, leaving its trail full of lived or memorialized history. This, attached to the present, corresponds to an elaborate compilation of material and immaterial symbols which, however distant they may be from the eyes, hands and other senses, are responsible for endowing our lives with meanings and values, since we are the result of our experiences lived over time (TUAN, 2013).

The domain discussed here, referring to devotion, benevolence, attachment, closeness and the nostalgic neighborhood ties that prevailed in memorial times on Ilha de Guaratiba, make up one of the past symbolic mosaics disseminated by members of the community. In addition to the representative elements highlighted on the previous pages, other events and characteristics of yesteryear continue to permeate the geographic soul of Guaratibanos who miss those memorable times. In this vein, let's consider another rich account by resident Benedito:

> The headquarters of the Ilha Futebol Clube is one of the town's most emblematic symbols. The "Clube do Ilha" is the element that best represents the culture of the place and the way of life of the Guaratibanos. How vibrant the Sundays at the Ilha Futebol Clube were! And the parties at Clube do Ilha! These were memorable experiences that marked my generation. The soccer field and the clubhouse were the hotspots for the locals. I also get very attached to the past when I remember the samba in the "quadra" and the parades we took part in every year. The carnival here on Ilha de Guaratiba was incredible back then (Dito - salesman - 50 years old).

The built symbols of Ilha de Guaratiba, alluded to in the above account, are exactly those that currently remain in a condition of resplendence in the hearts of people like the resident Dito, considering the glamor and luminosity they welcomed in other times. This is the case of the rehearsal court of the Grêmio Recreativo Escola de Samba União de Guaratiba, the headquarters of the Ilha Futebol Clube and its field, the Hélio Pantaleâo de Mello stadium. Until the 1980s, these artifacts safeguarded enthusiastic popular manifestations in the midst of the greatest passions of Brazilians in general and Guaratibanos in particular, namely carnival and soccer.

The street carnival in Ilha de Guaratiba has always been one of the liveliest in the area. As far back as the 1960s, the carnival festivities promoted large gatherings in the town's main thoroughfare, Largo da Ilha, where revelers from various localities converged. This innate quality of the Guaratiba people in promoting such revelry culminated in the emergence of the Unidos da Ilha de Guaratiba carnival block, created in 1977. The interviewee, Mr. Benedito, known locally as Dito, has always been an enthusiast of this group, being one of the main samba singers in the pre-carnival periods. In its early years, the bloco would parade in Campo Grande, during festivities organized by the Administrative Region, which also brought together other carnival blocos from the surrounding area. From 1980 onwards, however, the group was registered with RIOTUR and became part of the city's Carnival Blocks Federation and also received funding from this organization. The bloco began its professionalization by joining group 8 and parading in Padre Miguel. As a carnival block, Unidos da

Ilha de Guaratiba reached its peak in 1992. The following year, the Board of Directors opted to transform the bloco into a samba school, as it was now entitled to do so. From then on, G.R.E.S. Uniâo de Guaratiba never won a single carnival and fell into decline. Its rehearsal court, once crowded with revellers, was recently taken over by a timber company and turned into a large lumber yard (FERNANDES, 2010).

Soccer has also left its mark on the place. An example of this is the memorial reference to the Ilha Futebol Clube pitch and its headquarters, where Guaratibanos like Mr. Dito lived through remarkable moments, experiences that elevated these artifacts, now deteriorated, to the status of symbols of the place. This points to the permanent and intimate nature of the symbolic charge of places (TUAN, 2013; MELLO, 2003).

Ilha Futebol Clube was founded in 1948 by one of the most influential Guaratibanos of the time, Mr. Hélio Pantaleao de Mello. Its field, named after its pioneering entrepreneur, and its clubhouse, the Clube do Ilha, were recreational options for the locality for decades. That said, it's worth noting that another part of Benedito's account, not transcribed above, reveals other nuances of the relationship life of this club. In this context, despite never becoming professional, the Ilha team had organized supporters until the 1980s. The championships, tournaments and matches promoted by the club were very popular and hotly contested, with the pitch always packed with fans and admirers of the team. Daytime activities on the pitch were intense, with training sessions and matches for the various youth divisions: children's, youth, second division and first division. On Friday, Saturday and Sunday nights, the fun was to be had at the lively parties and dances held at the clubhouse.

The remembrance of the samba school's court and the momesco triduo, as well as the Ilha field and soccer as symbols, reflect the significance of these elements for a person who lived there during key moments in their life. The identification with these artifacts, forged through experiences, transformed them into enduring symbols.

With regard to the symbolic universe of an island of yesteryear, immortalized in memory, let's see what the following story has in store for us:

Because for decades it was our only means of transportation, the cable car was everything to us. I miss the streetcar very much. Until the 1960s, when it stopped running, it was a source of great pride for Guaratibanos. After all, very few people were able to enjoy this form of transportation at the time. The Roberto Burle Marx site is another symbol of Ilha de Guaratiba, as is the Sâo Salvador do Mundo chapel (Mazinho - retired - 66 years old).

The streetcar was an important element in the process of urban expansion in Rio de Janeiro (WEID, 1997). For Abreu (2008), the two elements responsible for the city's expansion were the trains (running on the tracks from 1858) and the streetcars, running after 1868. Trains and streetcars began to act together in terms of the connection and dance of places, if we want to use the conceptual ideas of David Seamon (1980) and Paulo Mauricio Rangel (2013) in the Portuguese version published by the magazine Geograficidade and as recorded in the work of Mauricio Abreu (2008) specified below. In this sense, we can say that the streetcars connected the city's most remote neighborhoods to the stations belonging to the Dom Pedro II Railway, inaugurated in 1858 and extended to Santa Cruz in 1878. The railway line that connected Ilha de Guaratiba to the Campo Grande railway station was inaugurated in 1924. With a length of approximately 20 kilometers, this route, managed by the municipality's now-defunct rural transport service, was deactivated in 1967, when its tracks were covered by the asphalt of the current Estrada da Ilha. However, decades after its extinction, the Ilha cable car (figure 7) is still remembered and revered as one of the past symbols of the place by resident Mazinho. The streetcars were part of the choreography of daily life, transporting the working masses, students and ordinary people to and from work. Road fever and the introduction of the automobile industry in the country in 1957 led to the removal of streetcar tracks in Rio de Janeiro, then the state of Guanabara, during the Lacerda government (1960-1965) and the administration of Negrâo de Lima (1965-1970).

Figure 7 - Bonde da Ilha in 1967, the last year of its circulation on line 11, which connected the town to the Campo Grande train station.

Source: PORTAL Guaratiba. Available at: <http://www.portalguaratiba.com.br>. Accessed on: November 20, 2015.

Other cultural elements emerge in Mazinho's testimony. He elevates Sitio Roberto Burle Marx and the Sâo Salvador do Mundo Church to the status of symbols of the place. As mentioned in the previous chapter, Burle Marx bought Sitio Santo Antônio da Bica in 1949. On the site there was an old farmhouse and a small 17th century chapel (figure 8) dedicated to Saint Anthony. Burle Marx restored both buildings and brought his plant collection to the site, which has continued to grow ever since. In 1973, the landscaper moved permanently to the site, where he died on June 4, 1994. Earlier, in 1985, the renowned Burle Marx donated his site to the Brazilian government, which administers it through the National Historical and Artistic Heritage Institute (IPHAN). Despite the historical and scientific importance of the Roberto Burle Marx Site, which attracts a considerable influx of visitors every day, its allusion as a symbol is the result of the role played by its founder in the locality. Burle Marx died more than twenty years ago. His influence, however, lives on, both in the physical layout of the place and in the minds of its residents.

igure 8 - Santo Antônio da Bica church, built in 1690, now part of the Roberto Burle Marx site.

Source: Roberto Burle Marx SiTIO. Available at: <http://sitioburlemarx.blogspot.com.br>. Accessed on: November 20, 2015.

Outside the grounds of the Sitio do Doutor Honoris Causa Roberto Burle Marx (a title awarded by the Royal Academy of Fine Arts in The Hague, Holland, and the Royal College of Art in London, England, both in 1982), another cultural symbol has fascinated residents and visitors to Guaratiba Island for centuries: the Igreja Sao Salvador do Mundo or Igreja Matriz de Guaratiba (figure 9). Built on top of a hill in 1676, the church still retains its original architecture (SANTOS, 1965; FRIDMAN, 1999). Because of its beauty and history, the Igreja Matriz also deserves to be mentioned by Guaratibanos like Mazinho, who elevates this church to the status of symbol of the place.

Figure 9 - Church of the Savior of the World: Guaratiba Mother Church

Source: REDONDO, 2012.

Imbued with the desire to translate this island of symbols, we considered the stories of Guaratibanos of different ages, such as that of young Vitória. Let's see what she has to say:

The Grota Funda Tunnel is the ultimate symbol and the greatest driver of this whole process of change that our place is undergoing.

passing. Before it was built, there was still the mountain that separated us from the rest of the city. It was common to cross the massif in an hour. This was due to the constant traffic jams, made worse by the fact that there was only one dual carriageway. This was the only way to Recreio, Barra and the South Zone. People thought twice before coming to live here. Because of the Serra da Grota Funda, the journey was long and tiring. Now, only two minutes separate us from Recreio. Since 2012, when the tunnel was inaugurated, Ilha de Guaratiba has been in the media spotlight. The place was finally connected to the city. The Serra da Grota Funda reminded us of something kind of mystical. It gave us the impression that when we left the city, we entered another world. This is no longer the case. We enter the tunnel and, two minutes later, we're already on Ilha de Guaratiba. Today, I no longer see so much difference between the place and the rest of the city. After the opening of the tunnel, in practice, our neighborhood ceased to be rural. Before, we were still hidden and protected from the city by the Serra da Grota Funda. Now, however, we are exposed. We are part of the city that has grown in this direction. The tunnel was built to continue the process of urbanization that is taking place in Barra da Tijuca and Recreio dos Bandeirantes. The Baixada de Guaratiba is the only large tract of land in the city that is still devoted to nature. There's still a lot of empty space here, and that's what the developers are looking for (Vitória - student - 21 years old).

In relation to the current context of the place, it is increasingly common for residents of Ilha de Guaratiba to choose a particular symbol, namely the Grota Funda Tunnel. This is a trend, especially among younger residents like Vitória.

The Tùnel Vice-Presidente José Alencar (Tùnel da Grota Funda), an underground link under the Serra da Grota Funda (Serra Geral de Guaratiba), connects the Guaratiba lowlands (Guaratiba Island) to the Recreio dos Bandeirantes neighborhood. The proposal to build this tunnel had been debated since the 1950s. However, the project, announced by various governments, always ran into a number of environmental/political/economic/administrative problems and it took decades before it got off the ground. With the implementation of the road corridor known as "Transoeste", work began on drilling the tunnel on September 14, 2010. It is worth noting that this project is directly linked to the two major sporting mega-events that the city is hosting, the 2014 World Cup and the 2016 Olympics. This justifies the presence of the president of the BOC (Brazilian Olympic Committee, Artur Nuzman) and the president of the CBF (Brazilian Football Confederation) at the time, José Maria Marin, as well as the city's mayor, Eduardo Paes, on the day construction began and, after a while, on the occasion of its inauguration.

The drilling of the tunnel's 1112-meter length, in two galleries, took eight months to complete. On June 6, 2012, the tunnel was finally inaugurated, considerably reducing the distance and time spent between Ilha de Guaratiba and Recreio dos Bandeirantes.

The physical characteristics of the city, characterized by numerous elevations, have always represented difficulties imposed by the environment on the urban expansion of Rio de Janeiro. In this sense, it is worth repeating that the drilling of tunnels, since 1887, has been a pressing need and symbolizes the connection of an area, previously consecrated to nature, to the city context (ABREU, 2008; CARVALHO, 2004).

Figure 10 - Grota Funda Tunnel: access to Ilha de Guaratiba on the left and Recreio dos Bandeirantes on the right

Source: RIO 2016. Available at: <http://www.rio2016.com>. Accessed on: November 20, 2015.

Considering both the story of the young university student Vitória and the analysis of experts on the urban question, we can emphasize that - in the post-tunnel period - the urban is inexorably moving westwards. In this sense, the city grows towards Guaratiba Island, with the Grota Funda tunnel as one of the drivers of this urbanizing march and a symbol of this change (LESSA, 2001; CARVALHO, 2004; REDONDO, 2012; JANOT, 2013).

As well as putting Ilha de Guaratiba on the urban map of Rio de Janeiro (BERTA, 2012), the Grota Funda tunnel also represented an unprecedented increase in the value of land and real estate in the area. This positive and vertiginous evaluation has contributed to the promotion of a series of value factors that we would like to address in the final chapter of this research.

6 ENHANCING THE VALUE OF SPACE BY PROMOTING

"A Ilha é seca, mas é cercada de amor" (Quinzinho do Impèrio - lyrics from Unidos da Ilha de Guaratiba's samba-enredo in the 1980s).

Commonly, the concepts of space and place express - metaphorically and respectively - the notions of darkness and luminosity (MELLO, 2000). In this way, far from the positivist and neopositivist impositions, there is no pre-established condition for opaque spaces, immersed in darkness, to reach the level of place through their illumination or clarity. Paraphrasing Mello (2003), we can point out that a place can lose or gain this status depending on the darkness or brightness with which it becomes overshadowed or even illuminated over time. According to Tuan (2013, p.199), "place is any stable object that captures our attention". However, the same thinker ponders: "many places, highly significant to certain individuals and groups, have little visual notoriety" (TUAN, 2013, p. 200). However opaque or visible they may be, certain objects or places admired by one person may not be noticed by another (TUAN, 2013).

Place is, par excellence, a reservoir of memories and dreams (TUAN, 2013). In these circumstances, visual notoriety alone is not an instrument for certain spaces to become places. A deep sense of place results from a combination of historical, geographical, cultural, economic, existential, subjective, intersubjective, immaterial and visual factors and values (YAZIGI, 2003). In terms of the line of thought employed in this research, visibility is not only linked to human constructions and elements of nature that give visual values to certain artifacts. However, there is no denying that these natural and cultural attributes, in many cases, represent factors in the transformation of certain locations which, as a result, can undergo qualitative or non-qualitative changes, depending on the different perspectives of those who experience them. With regard to the different viewpoints of individuals in relation to visibility, Tuan stresses that most places are not deliberate creations, as they are built to satisfy practical needs. In this context, they acquire visibility and meaning for both locals and outsiders (TUAN, 2013).

The boundaries of a place are fluid and existentially demarcated by its residents and inhabitants. On the other hand, the same perimeter represents a mystery to outsiders, visitors and newcomers. This niche of "ambiguities, topophilic feelings, fears and people's philosophical way of acting" (MELLO, 2000, p. 129), can forge spaces and places, respectively, through the valorization and valuation of undifferentiated places, spaces which until then had been treated with relative indifference even by their insiders (FERNANDES, 2006). In this sense, visibility, related to natural and built landscape attributes, becomes a significant element, both for insiders - who begin to give more value to the place they live in after its valorization - and for the new residents who see these peculiarities as indispensable factors for the valorization of a space that for them was undifferentiated, thus making this lived universe a home or place.

Embracing this philosophy about the places of the men and women focused on in our investigation, we understand that the process of real estate development that has given visibility to the residential areas of Ilha de Guaratiba has promoted significant changes in the existential way of life of the local residents. It is these existential metamorphoses that flourish in the midst of this urbanization march that will be discussed below, starting with the role played by valuation in the phenomenon of transforming space into place.

6.1 From the valorization of space to the valorization of place

Valuation - here understood as absolute value or objective value, price or economic value attributed to a given object, asset or area - when related to the forms of housing directed by the real estate market, can become enlightening to elucidate the dynamics involving the phenomenon of place-making. In this sense, visibility, understood as the visual character of a given locality, may be the main factor responsible for that area achieving a certain status related to its real estate value, as is the case with most neighborhoods with significant income strata (TUAN, 2013).

In the US context, even in the first half of the last century, Walter Firey (1945; 2013) was already looking at the network of relationships between the symbolic and sentimental charge of places (valuation) and their residential attractiveness for the economically privileged classes (valuation). In Firey's view, certain places should not be attributed solely to the economic variable, since they are also endowed with a kind of "sentimental imprint". Among the examples cited in Walter Firey's study is the case of an upper-income residential neighborhood known as Beacon Hill, located in central Boston. In the study in question, Firey points out that the growing economic value of the Beacon Hill neighborhood is due above all to its symbolic charge and the sentimental associations built up over its century-long history. "In Bacon Hill, a series of aesthetic, historical and family feelings have been spatially articulated" (FIREY, 2013, P. 23).

Regarding the link between the real estate development that is taking place on Ilha de Guaratiba (FERNANDES, 2006) and its "spatial representations" (FIREY, 2013, p. 22), let's consider another statement by Mrs. Dalva:

> From the moment they started to see important people: Wando, Victor Fazzano, Ernani Moraes etc. buying a place or a house on Ilha de Guaratiba, the residents who didn't like the place started to wonder: why are these guys coming here? From that moment on, the younger Guaratibanos realized that their place had something special. Ilha de Guaratiba began to have value for these people. The rise in real estate prices brought many changes to the area. The place grew and many new people moved to Ilha de Guaratiba. This increase in people and urban activities has made young people value the place more. Twenty years ago, young people thought they lived at the end of the world. They didn't like Ilha de Guaratiba. Today, however, these young people have matured and started to value the place more. What's more, environmental issues are very much in vogue at the moment. This has contributed to people (young and old) becoming more attached to nature. In this context, the development of Ilha de Guaratiba is causing concern among its residents, who now value the bucolic nature of their place more. If people from noble and traditional neighborhoods are coming to live here, it's because the place has value, isn't it? A few years ago, young people here wanted to be in the places where a considerable number of the new residents come from. In addition, the increase in urban violence has made young Guaratibanos value their neighborhood more. The appreciation that has come from other people, from other places, has certainly contributed to us - Guaratibanos - "valuing" our place even more. Today, even our young people are more attached to the place (Dalva - housewife - 66 years old).

There is a maxim, commonly used in common sense, according to which "people only value things after they have lost them". This axiom reveals a recurring human tendency, since the absence of an essential good, a relevant feature or artifact, or a loved or dear person, in many cases leads us to recognize the importance of their existence or presence. In this sense, presence and absence are directly related, since the absence of a loved one, for example, can be largely responsible for their presence in our minds and hearts (LEFEBVRE, 1983). Not only absence, but also the fear of losing something indispensable, can trigger a series of changes in the posture and existential way of life of certain individuals and social groups (BUTTIMER, 2015).

The question here is simple and consists of unraveling the following question: can the process of (economic) valorization that is taking place on Ilha de Guaratiba influence the way residents experience the place? Could it be that after being economically valued, the place became more "valued" by its residents?

That said, we can emphasize that the restlessness of the Guaratiba residents, caused by fear or suspicion of the rivalry represented by the presence of new residents in their place, may have been the preamble to a new relationship with their lived universe, as Dona Dalva's account points out. Looking at her testimony, we can see that, in the specific case of Ilha de Guaratiba, real estate development didn't just bring about spatial changes. For different reasons, the *addition of* new homes and residents has triggered a change in the attitude of younger Guaratibanos who, following this development, have come to "value" the place they used to treat with indifference. In this respect, considering the above-mentioned statement, we can emphasize that this new relationship of the younger people can be understood as a true "existential metamorphosis", since the detachment from their former residential "space" has metamorphosed into affection for their current "place". With regard to the feelings pertinent to the changes brought about by speculation, appreciation and urbanization, Guaratibanos - jealous, perplexed and frightened to see their lived world being invaded by people and families from other neighbourhoods in the midst of this process of invasion-succession (CORRÊA, 2000) - begin to strengthen their ties with the lived universe even more in the condition of place.

Valuation - understood in terms of subjectivity, esteem, affective value, identity charge (HAESBAERT, 2004), references for the construction of spatial identities and a sense of belonging (SOUZA, 2004) - denotes the symbolic and/or philosophical value attributed to a given place, and is constructed from lived experiences. The transition from space to place mainly involves the affective and symbolic value of the spatial portion in which the individual is inserted. In many cases, however, the economic or other value conferred on an artifact, street or area can be a fundamental element in building bonds and a sense of belonging. In this way, we can infer that, in specific cases, the valuation of a given space can produce its valuation and consequent transformation into a place (FERNANDES, 2003; 2006; 2010).

Considering that the individual is not distinct from their place (COSGROVE,

2004; LOWENTHAL, 1982), as all humanistic geographers reiterate, we also conclude that events occurring in a certain spatial domain can influence, directly or indirectly, the lives of those who experience it. Spatial metamorphoses are channeled by human beings and can change the way they live and even the way they relate to their lived place. In these circumstances, the affective bond between the person and the place, forged in everyday life, can be weakened or strengthened, depending on how the changes are introjected.

By reading the testimony used and given by Dona Dalva on this investigative scale, we can see that, before speculation and real estate development, the place was treated with a certain indifference, especially by its younger residents. In a clear ethnocentric stance, they believed they lived at the "end of the world" or "where Judas lost his boots", wishing to migrate to brighter places, where the attributes of urbanity would provide them with a dynamic life. That said, it should be remembered that ethnocentrism takes on the form of valuing the place and the person and, on the other hand, the internalization of negative aspects leads to the depreciation of the place lived in and of one's own self-esteem (TUAN, 1980; 2013; MELLO, 2000). For them, Ilha de Guaratiba was nothing more than an undifferentiated space, since it was not valued by them (TUAN, 2013; FERNANDES, 2003; 2006; 2010). However, after the (economic) valorization that gave the place relevance, these *insiders*, noticing their lived universe being taken over by *outsiders*, began to change their minds, developing a different feeling towards it. As a result of this phenomenon, young Guaratibanos begin to give more value to their lived world. This value, which emerged from appreciation, triggered a change in attitude among the residents, who began to nurture feelings for their lived universe, based on a new relationship (FERNANDES 2003; 2006; 2010). The indifference, contempt, disdain, rejection, disregard, disinterest, apathy and insensitivity of yesteryear are now replaced by admiration, pride, affection, sympathy, satisfaction, love and other valued feelings responsible for verbal reports that both express intimate relationships with the place and demonstrate its expression as such. However, before trying to translate the topophilic feelings of Guaratibanos - manifested from new experiences with their lived world - we will then try to capture the anguish of Guaratibanos who are fearful of a supposed expulsion - due to economic valuation - from their lived universe.

6.2 From the promotion of place to the expulsion of the lived universe

Despite the certain indifference with which Ilha de Guaratiba was treated by its residents - a phenomenon highlighted in the previous topic - and through the series of events that brought about the spatial and existential changes also dissected in this research, intimate experiences emerged in the midst of this process that culminated in the acquisition of a place par excellence. After helping to promote an undifferentiated space to the status of a place for its residents, real estate development continued to follow its course. However, after the inauguration of the Grota Funda Tunnel, the exacerbation of this process culminated in a pressing concern on the part of Guaratibanos: the fear that the changes underway will expel them from their lived universe. But what do the residents of Ilha de Guaratiba have to say about this concern?

If they really do change the Urban Structuring Project (PEU) for the town, the only way out for most of the residents would be to sell their land and find somewhere else to live. In fact, the town hall is already anticipating the continuation of the urbanization process that is taking place there, driving the Guaratibanos out of their place. Barra is already saturated, Recreio is even suffering from favelization. The city has nowhere else to grow. How ironic: the place appreciates in value, people from outside start coming here and then we're kicked out of our own home. We will reach a point where many residents will no longer be able to pay the IPTU. This is a process that has already happened in places like Vargem Grande and Recreio dos Bandeirantes. A few years ago, many people left those places and came here because of the higher tax burden they had to pay. The same tends to happen here. In fact, the replacement of the Rural Land Tax (ITR) by the Urban Land Tax (IPTU) is already driving some families out of the area. Many people have sold their property at low prices out of fear or to avoid the urban tax. I am representing a family that used to pay R$ 80.00 a year in ITR and now has to pay R$ 40,000.00 in IPTU. The property was bought a few years ago from a family who sold it in order to escape this exorbitant tax burden. My aim is to have only the property tax levied, as required by current legislation, and not the land tax, as has happened in this specific case. With the replacement of the ITR by the IPTU, the expulsion of Guaratibanos is almost certain. Families from Estrada das Tachas and Caminho do Morgado, humble people who paid ITR on areas of 10,000 m^2 , are having to sell their properties. Guaratibanos who have a standard area, according to the PEU in force (10,000 m2), won't be able to stay here because of the IPTU. Everything has its price. The government wouldn't build the Grota Funda tunnel and the Transoeste highway for free. The government itself is going to expel Guaratibanos from here to make room for a population that can pay for these "improvements". In my opinion, all this is a government strategy to destroy our place (Tadeu Garrido - lawyer - 49 years old).

Ilha de Guaratiba is still subject to the delimitation established in the 1970s, which only takes into account the rural characteristics of the time, with a minimum plot size of ten thousand square meters (BERTA, 2012). In this context, landowners in the area were charged Rural Land Tax (ITR) and were exempt from Urban Land Tax (IPTU). However, since the last IBGE census - in 2010 - the population of the city of Rio de Janeiro has been considered 100% urban. Using this subterfuge, a few months later, the city council changed the legislation and, likewise, began to characterize the municipal territory as entirely urban according to the 2011 master plan (complementary law 111 enacted on 01/02/2011).

According to Berta (2012), once Ilha de Guaratiba was considered an urban area by the municipality, the city council had to work on the necessary Urban Structuring Project (PEU), in order to define the neighborhood's urban standards and ensure occupation that would not make it impossible to maintain the natural environment and its traditional production of ornamental plants. However - the PEU, which could redefine the new urban patterns of the area and regulate the form of occupation, the minimum size of plots and the maximum number of floors - has not yet been defined.

In theory, the Urban Structure Project is a set of rules guided by policies and actions defined to guide the physical-urban development of a group of neighboring districts with similar characteristics. In practice, however, the PEU is also used to define the IPTU that will be charged. Following this redefinition of the area in question, a 20,000 or 30,000 square meter plot of land that used to be used by a family of farmers who made a living from the land, under which the inexpensive federal tax (ITR) had been levied, now faces an annual IPTU debt of 30,000 reais (BERTA, 2012).

The issues highlighted in the last three paragraphs converge with what lawyer Tadeu Garrido said in his testimony about the apprehension of the residents of Ilha de Guaratiba who - in the midst of the spatial, administrative and tax changes brought about by the inexorable march of urbanization - express their dilemmas, anxieties, worries and fears in the face of a possible expulsion from their lived universe. This apprehension on the part of the Guaratibanos became more acute in the post-tunnel period, when the place gained visibility. In addition, real estate speculation has further boosted real estate and land values, a factor that has intensified the "march west" in recent years.

As we can see, there are many concerns among the residents of Ilha de Guaratiba in relation to modern events: accelerated population growth, construction of multi-family buildings, degradation of natural environments, making rural production (of ornamental plants) unfeasible, etc. Looking at Tadeu Garrido's account, however, the main concern of long-time Guaratibanos becomes clear: the pure and simple change from federal taxation (ITR) to municipal taxation (IPTU). This fear is justified by the fact that many residents live in areas of at least 10,000 m^2 . As most of these properties are productive, residents question the fact that the IPTU can only be levied on the area built on (building), and not on the whole property (land), a fact that has occurred on some properties, as the lawyer in question pointed out.

Due to the lack of specific legislation that allows for differentiated taxation for each type of property, Ilha de Guaratiba has different realities: there are properties that, because they represent possessions, do not pay any tax; there are others that, because they prove to be productive, pay only the ITR; there are those that pay the IPTU and there are also some where there is double taxation (ITR and IPTU).

In the process of spatial change that is taking place on Ilha de Guaratiba, real estate speculation can be considered a two-way street. By promoting real estate appreciation, this speculative practice has helped residents to "value" their former space, elevating it to the status of a place (FERNANDES, 2006). Nowadays, however, the continuity of the process of valorization - accompanied by its nuances - can contribute to Guaratibanos being "expelled" from their lived universe. This is the contradiction: the valorization that helped elevate a once undifferentiated space to the status of a place, through the residents' appreciation of their experienced ground, can today be responsible for separating Guaratibanos from their place.

We can better understand the anguish, protests, fears and fears of Guaratibanos when we look at the relationship they have built with their neighborhood over time. We can then see that this attachment to place is the result of clashes, joys, disappointments and other affective experiences - producing a niche of experiences and vacillating feelings.

6.3 **Topophilia: intimate experiences with place**

Let's take a look, in the lines below, at another verbal account of a Guaratibano in love with his place:

> Ilha de Guaratiba is my base, my home. I'm very grateful to my good God for allowing me to meet him. In this place I have met wonderful people, as well as an environment with which I have a very strong and intimate relationship. I can't imagine living anywhere else (Paulinho - civil servant - 45 years old).

Paulinho, as public servant Paulo César is better known, was born and raised near Ilha de Guaratiba, building an affinity with the place through soccer and the local parties. However, this relationship grew stronger when Paulinho married a local woman and moved to Ilha de Guaratiba. Paulo César's passion for the place is such that he attributes the fact that he met it to the Creator, as if it were a person, a loved one. More than that.

Transcendental, his unbridled love for the place is on a level of devotion. Also noteworthy is the resident's attachment to the people of the place, proving that "the more ties there are, the stronger the emotional bond" (TUAN, 1983, p.175).

Let's now take a look at another resident of Ilha de Guaratiba's statement about his attachment to the place:

> Ilha de Guaratiba has always been my nest, the best place to stay. When I was younger, I had a girlfriend who kept asking me to leave this place, but I always replied: you can't leave it here. Where else can you find a starry sky like this? And that cool breeze? This different climate? This vegetation? These birds around us? These flowers and plants? This contact with life? Here I have everything I need to live (Marcelo Paes Costa - landscaper - 48 years old).

Chel, the codename of landscaper Marcelo, is one of those Guaratibanos who feels a deep sense of pride at being a native of the place where he was also raised. The topophilic ties between Marcelo and Ilha de Guaratiba were built up over the course of his experiences living in the aforementioned spatial portion of bem querência, an experience responsible for a strong attachment and identification with the territorial base. Unlike Paulinho, who stresses the importance of people and interpersonal relationships as the main stimulus for his identification with the place, Chel prefers to exalt the natural and scenic attributes of the place, denoting that these are the ones most responsible for the link between him and his lived world.

Let's take a look at the words of another resident regarding his splendid love for Guaratiba Island:

> This is where I feel good, at home and complete. I was born here, my wife was born here, my parents were born here. My pianos and dreams are always about something being built here. I've never considered going elsewhere, because this is where I want to build my life. The emotional issue weighs heavily on this kind of decision. In addition, I would like to contribute in some way to Ilha de Guaratiba for everything it has given me since childhood. Because it's so special to me, I really want this place to do well, and whatever I can do to help... I'm right here (Marion - 41 years old - employee of Sitio Roberto Burle Marx).

The propositions that "places are loved ones deserving of consideration" (MELLO, p. 50) and that in the "symbiotic relationship between men and the environment, places must be considered people and people as places" (POCOCK, 1981, p.337) are alluded to in the above account. In it, the landscaper Marlon emphasizes wanting the good of Ilha de Guaratiba as if it were a person deserving of his consideration and esteem for everything it has given him throughout his life. In his statement that he feels complete in this environment, Marlon is moving towards the humanistic premise that "place is a part of the earth's surface in which man completes himself" (MELLO, 1991, p.50).

In Tuan (1980), the term topophilia is defined in different ways, including "the manifestations of human love for place" (p.106), "the affective ties of human beings with the material environment" (p.107), in short, this term is always associated with the feeling of the individual towards the place (p. 129). Affiliation, attachment, affection, love, identification and other qualitative feelings people have towards a certain environment, however, are established through intimate experiences, in which this physical environment is elevated to the level of place (TUAN, 1883). The intense feeling of love for place, however, is often deepened by simple events, since

> Intimate experiences, when not exalted, go unnoticed. At the time, we don't say "this is it", as we do when admiring objects of notorious or recognized beauty. It's only when we reflect that we recognize its value. At the time, we are not aware of any drama; we don't know that the seeds of a lasting feeling have just been planted. Simple events can, over time, develop into a deep feeling for the place (TUAN, 1983, p.158).

Intimate experiences, therefore, are not necessarily linked to major events, but rather to everyday events which, despite their apparent simplicity, are also responsible for people's attachment to their place. Being personal and individualized, these intimate experiences are difficult to express (TUAN, 1983), translate and decode. When revealed through spoken or written accounts, however, these experiences manifest beautiful stories of individuals' and social groups' love for their place, where it is situated by people as their own extension. In this sense, in the midst of introjections, the individual cannot distinguish themselves from their place, in which case there is no separation between subject and object (COSGROVE, 2004; LOWENTHAL, 1982; RELPH, 1976; SCHULZ, 1979). With regard to the pertinent simplicity of the experiences that can lead individuals to become attached to a place, the following statement may prove enlightening:

> This place formed me and allowed my subjectivity to blossom. It influenced me in everything. It was where I experienced moments of loneliness. Today, however, I have a huge bond with it. It's where I intend to raise my children. Here I learned to value simplicity, the people who live off the land and habits such as going to the mangrove to pick guaiamu, going to the fields to pick manioc, as well as other simple values such as bringing people together to eat, drink? This sense of community, of being able to greet and talk to people in the street. This closeness to people I learned here, where I feel at home (Kleber - sociologist - 39 years old).

With regard to the feeling of belonging that makes Kleber feel at home in this place, Schutz (1979, p. 291) argues that "feeling at home is an expression of the highest degree of familiarity and intimacy" and means, among other things, customs, values, personal habits, traditions, a peculiar way of life, made up of small, important and cherished elements (SCHUTZ, 1979). In the specific case of the sociologist Kleber, we can understand that it was precisely these day-to-day peculiarities that exerted the greatest influences, both on his individuality and on the strong bond he built with the place. Coming from Tijuca, where he came from with his family as a child in 1982, Kleber had to overcome moments of loneliness until he adapted to his new surroundings. However, the initial discomfort was replaced over time by an irresistible attachment to his new "home". This episode shows that love for a place, as well as the feeling of belonging and affiliation attached to it, does not lie in the fact of being its "legitimate son", but is mainly associated with the emotional experiences lived in a certain space which, through these affective ties, becomes a place. In this way, in the same way that a parent's love for their adopted child is linked to experiences and not to blood ties, an individual's love for their place is not subordinated to the fact of having been born there, but rather to experiences, contacts, relationships, coexistence, learning and values in the midst of relatives, friends, acquaintances, feelings and so on, making up a whole of strangeness, adherence and belonging (MELLO, 2000). Through intense intimate experiences, such as those lived by Kleber on Ilha de Guaratiba, the place can become the center (navel) of the world for certain individuals and social groups. This feeling has positive elements of valuing the place lived, the best in the world, or the center, where everything converges, as in the case of the feeling conveyed by the sociologist Kleber. It can, however, have a negative connotation, of rejection, aversion, of the ugly, bad, distant space, in relation to the end of the world, where "Judas lost his boots" and so on (TUAN, 1980; 2012; MELLO, 1991; 2000).

6.4 **Ethnocentrism: the place as the center / navel of the world**

For many people who have heard of the place in question, it represents nothing more than an undifferentiated peripheral area, far from everything and everyone. On the other hand, since the passion they experience doesn't share with dull, distant thoughts, there are Guaratibanos who place their place at the center of their world. With regard to the ethnocentric stance, the account below is representative:

The worst period of my life was when I had to leave Ilha de Guaratiba. Because of my son, I had to move to Jacarepagua, where I lived for 16 years. The fact that I lived in another neighborhood, however, ignited my love for this place even more. I used to come here practically every day, because here was my mother, my work, my things, my life. I've always been attached to this place, and the first chance I got, I rushed back here and I've never left, and I don't intend to.

When I go on long trips, for the first two days I'm still eager to get to know things and places. However, after the third day, I feel a crazy urge to go back to where I came from. The midwife who helped me give birth must have buried my belly button in that piece of ground. Perhaps this explains why I enjoy coming home so much when I'm traveling. When I lived in Jacarepagua, I didn't feel much like going home, but I always want to go back and I'm happy to stay.

On holidays or weekends when there's no one in my house, taking a walk inside is the most satisfying thing. It's very difficult to talk about my connection with Ilha de Guaratiba. Perhaps the phrase that comes closest to fitting is "a relationship of flesh and blood". In any case, it's too little to describe how I really feel. Words fail me... (Enanir de Souza - businessman - 75 years old).

Since the place, in its condition of stability and confinement, reminds us of a feeling of an intimate and humanized home and a niche of protection and coexistence built up in the daily grind, obviously, moving away from this point of support and well-wishing is a mismatch that produces disenchantment and discontent. This was the feeling Evanir de Souza had when he had to move away from his arena of struggle and affection, which he explored with singular ease (TUAN, 1983, 1998; MELLO, 1991; 2000). This businessman makes a point of describing how much he misses his place when he is traveling. In this sense, his account is in line with the phenomenological postulates of Schutz (1979) when he discusses some issues related to returning home.

What, however, should "home" mean? "Home is where you leave," says the poet. "Home is the place to which a man intends to return when he is far away," says the jurist. Home is the starting point as well as the end point. It's the zero point of the coordinate system we assign to the world in order to move within it. Geographically, "home" means a certain place on the surface of the Earth. Where I happen to be is my "domicile"; where I intend to stay is my "residence"; where I come from and where I want to return to is my "home". However, home is not just the place - my house, my room, my garden, my city - but everything it symbolizes. The symbolic nature of the notion of "home" is emotionally evocative and difficult to describe. Home means different things to different people. It means, of course, the parental home, the mother tongue, family, love, friends; it means a beloved landscape, "songs my mother taught me", food prepared in a certain way, familiar things for everyday use, customs, personal habits - in short, a peculiar way of life, made up of small, important and cherished elements (SCHUTZ, 1979, p. 290-291).

At the end of her story, Evanir de Souza is still looking for a plausible answer to justify the deep attachment and appreciation she feels for her place, turning to an old habit of burying the external navel of children. This custom, quite common at a time when most births were carried out by midwives in their own homes, symbolized the attachment of the child who had just been born to the place where they had first seen the light. After a beautiful speech that showed a deep appreciation for her home base, Evanir do Souza surprised us by saying that her relationship with Ilha de Guaratiba was indescribable. As we know, intimate experiences, which are responsible for placing an affectively demarcated area at the center of certain people's worlds, are really difficult to communicate. However beautiful they may be, these evocations always fall short of the depth and experiences that take us to a center full of values (SCHUTZ, 1979).

According to Gomes (2007), the first fundamental characteristic of humanism to be taken up by geography is the unavoidable anthropocentric view, according to which man is the measure of all things. Furthermore, human beings, individually or in groups, tend to place their lived place at the center of the world. In this context, egocentrism and ethnocentrism become universal human traits (TUAN, 2012). Based on Tuan (1980), Mello (1991, p.202) describes ethnocentrism as a universal phenomenon of overvaluing the "center", "navel", "healthiest" or "best place in the world" and can also be understood as collective egocentrism. People from the "center" discriminate between "us" ("superior") and "them" ("of lesser value", "of inferior culture") by looking at the latter in a "blasé" way and sometimes with apathy, sarcasm or aggression.

The notion of the center is one of the most important cultural manifestations, since people tend to situate the place where they live as the most important and favorable, and as the center of their world. "All ancient peoples place themselves at the center of relationships and organize what they mean by the world in this reference". In this sense, the top is a related notion. "A whole religious and cartographic symbolism derives from this centrality, organizing the conception and geographical relations of these peoples" (MOREIRA, 2009, p. 67). According to Indian beliefs, Mount Meru stands at the center of the world. An Iranian myth believes that the sacred mountain of Elburs is located at the center of the Earth. The name of Mount Tabor in Palestine could mean "navel". Mount Garizim, in the central region of Palestine, undoubtedly enjoyed the prestige of a central place, as it was called the "navel of the Earth" (ELIADE, 2007). According to ancient tradition, which has been preserved in the region to this day, Palestine, in its condition as the highest country, being close to the summit of the cosmic mountain, would not have been covered by the flood. A rabbinical text states: "the land of Israel was not submerged by the flood". For Christians,

> the Golgotha was situated at the central point of the world, since it was the summit of the cosmic mountain and, at the same time, the place where Adam had been created and buried. Thus, the Savior's blood was shed on Adam's skull, buried precisely at the foot of the Cross, serving for his redemption. The belief that Golgotha was located at the center of the world is still preserved in the folklore of Eastern Christians (ELIADE, 2007, p.24).

From the beliefs we've mentioned, we can deduce that each eastern city was located at the center of the world. For some of these peoples, the highest point of the cosmic mountain was not only the highest point on Earth, but also the navel of the world, the point at which creation began. Some traditions explain the symbolism of the center in terms taken from embryology, according to which "the Divine Being created the world as an embryo. In the same way that the embryo was formed from the navel, God began to create the world from the navel onwards, and from there spread out in different directions" (ELIADE, 2007, p. 25). In the meantime, the world would have been created from Siâo and the universe conceived from a central point. The creation of man also took place at a central point, at the center of the world (ELIADE, 2007).

According to Mesopotamian tradition, man was formed in the "navel of the Earth". Thus, Paradise, where Adam was created from clay, is located at the center of the cosmos. Paradise was the navel of the Earth and, according to a Syrian tradition, it was established on a mountain higher than all the others. Adam was created in the center of the Earth (ELIADE, 2007). Based on the premise that creation was derived from a center, we can also assume that any place founded has its building in the center of the world of the individuals who established it as their home, shelter, refuge and dwelling place.

The chimera related to this type of centrality is, according to Tuan (1980; 2012), necessary for the maintenance of culture. For Tuan, when raw reality shatters the illusion that one's place is superior, culture itself may decline. Even knowing that they are not at the center of things in the literal sense, it is necessary for something of this faith to be present in small communities for them to prosper (TUAN, 1980). The progress that comes from the human belief that their lived universe is at the center of the world, however, is not necessarily tied to numbers, quantities and other objective values (valuation). Centrality, in this case, would be associated with a series of subjective values (valuation) based on the subjectivity or intersubjectivity of individuals and social groups. Since subjective value is a relationship between the subject who values and the object valued, attributing value to a place means not being indifferent to it. Non-indifference is the main characteristic of value, since we are not insensitive to a particular place that captures and radiates our affection and sympathy, because we are always influenced by experiences in the place we live. Valuing is a fundamentally human experience that lies at the heart of every choice in life (ARANHA and MARTINS, 1992). Among these choices is the choice of the spatial portion that distinguishes itself from all the others, not necessarily because of its form-context, but because of the experiences, the feeling of belonging, the affection and the other qualitative values that its Experiencers nurture for it, a valuation that will focus this once undifferentiated space at the navel of the world, as the center of the universe.

From a humanistic perspective, the concept of place, because it has no defined scale, becomes too diffuse since it can either designate a place or encompass the planet (TUAN, 1983). However, the same Tuan who takes up the maxim that defines Geography as "the study of the Earth as the home of people" (TUAN, 1991, p. 89), also points out that "topophilia sounds false when it is manifested by an extensive territory" (TUAN, 1980, p. 116). However, patriotic love, significant in its dimension, can contradict his elucidations. For the aforementioned geographer, topophilic feelings require "a compact size, reduced to the biological needs of the human being and the limited capacities of the senses". Furthermore, a person can identify more easily with an area if it appears to be a natural unit, small enough to be known personally (TUAN, 1980, p. 116-117). In this field, place is confused with the local sphere, and is also the locus of everyday life "responsible for human passions through communicative action and for various manifestations of spontaneity and creativity" (SANTOS,

2002, p. 322).

Ilha de Guaratiba is one of those small communities, proclaimed by its residents, a lived neighborhood that pulsating daily life has transformed into a center full of values, an inseparable whole made up of people, friends, acquaintances, relatives, a territorial base, evocations and other references, such as "songs that my mother taught me" (SCHUTZ, 1979) that allow its Experiencers the pleasant sensation of feeling at home or even immersed in the center-umbiglo-heart of the world (MELLO, 1991, 2000; TUAN, 1980, 1983, 1998), as the story below suggests.

Here I made my permanent home. The navels of my children I buried here. My grandchildren's navels will be buried here. I made my permanent home here and it is in our permanent home that a piece of our skin should be buried. There is already a piece of me buried here because I chose this place to live and build my life home, the house where I write and where I live the best moments of my life. Copacabana is where I live to be a doctor, in other words, it's where I work. Not here! This is the place where I complete myself. It's different from all the others because my relationship with it is different. Here I made and have the best friend in the world, who is Mrs. Seci; I have my godson, who I adore; it has that small place style that I have always liked, in other words, simple things that are in danger of disappearing with the arrival of the big real estate companies, being replaced by the metropolis (José Humberto Resende - doctor - 62 years old).

Unlike Guaratibano Evanir de Souza, who was born and raised on Ilha de Guaratiba, José Humberto Resende is just one of the many second-home owners in the area. The aforementioned doctor, who lives in Copacabana, bought a property in the area where, for decades, he has taken refuge during his vacations, weekends and vacations. These facts led Doctor Humberto, as he is better known, to develop an affection for the place. This inordinate affection is made explicit in his account, which reveals, among other things, his esteem, respect and concern for the place he has chosen as the center of his emotions. Consideration and appreciation are evidenced by the doctor's notorious attachment to the place he learned to love. The feeling of respect and concern is evident when José Humberto, in the midst of the spatial changes taking place in the place, fears that it will be profoundly altered, losing its main characteristics and taking on the atmosphere of the metropolis that surrounds it.

Dr. Humberto's final words shed light on the complex dynamics surrounding spatial transformations. Through his speech, we can conclude that the continuity of the process that led to his encounter with the place in the past now represents a risk to its maintenance, since the characteristics responsible for the succession of changes run the risk of overly metamorphosing the place.

As participants and members of this area, we have observed that the emergence and unfolding of this process has contributed decisively to the awakening of an affinity and fondness for the place that has led new and old Guaratibanos to understand the corner of the world in which they live as the center/umbiglood/heart of the world - matching the common tendency of people to consider the corner of the world in which they live as the only favorable one and their customs and habits as the quintessence of humanity (TUAN, 1980, 1986).

6.5 **Guaratiba Island: a place**

A "place is any stable object that captures our attention" (TUAN, 2013, p. 199). Its boundaries are existentially demarcated for its residents and experiencers. There are propositions that "places are loved ones worthy of consideration" and that in the "symbiotic relationship between men and the environment, places should be considered people and people as places" (POCOCK, 1981, p.337). Do the Guaratibanos feel this way? Let's consider the opinion of one of them:

> Ilha de Guaratiba was essential to my formation as a person. What I am today is the result of my experience there. I have always been proud to be a son of this land. This is where I was born, literally speaking. I was born in my house and my navel was buried in my backyard. Pipo and Ilha de Guaratiba: a visceral relationship. I've always loved my place. Over time, however, my love for it grew. As we mature, we come to value the truly important things more. If Ilha de Guaratiba were a person, it would certainly be the passion of my life. I am in love with Ilha de Guaratiba. I can't imagine living anywhere else. I consider myself a Guaratibano in body and soul. I'm planted on Ilha de Guaratiba. This place is my root. As the birthplace of my family, Ilha de Guaratiba gives me a very deep feeling. Ilha de Guaratiba deserves all my affection, respect and admiration. When I travel, I get excited on the first day. The next day, however, I start to feel homesick. By the third day, I'm itching to get back. Every trip has the same feeling. Not even when I'm far away can I get away from this place (Pipo - biologist - 49 years old).

What is a place? In theory, we have been trying to answer this difficult question since the very beginning of this research. In the second chapter, we looked exclusively at this concept in order to try to translate the different interpretations of its definition, as well as its multiple dimensions. But, in practice, what characteristics can be listed to elevate a locality to the status of a place? Is Ilha de Guaratiba a place? What makes this simple locality a place? These last questions cannot be answered on the basis of theories, since the answers are linked to the lived experiences of individuals and social groups in their lived universe. Only they can talk about their experiences, feelings and meanings built up over time on their experienced ground. Only the intensity of these experiences, as well as their interpretation, can define whether or not a place represents a place for its residents.

Based on the above testimony, we can consider Ilha de Guaratiba a place. As a representative of his community, Pipo, a resident who is passionate about his lived universe, contributes to this assertion through his account. Among the reasons and/or characteristics that guide the "locality" (RELPH, 2012) of Ilha de Guaratiba, the following are explicit in the resident's statement: the importance of the place for the formation of the individual as a person; the passion for the place as if it were a person; the deep affection for the place built up and strengthened over time; the feeling of belonging and rootedness to the place; the longing for the place when away from his "home" and the love for the place.

To close on the meaning of place, I'll leave you with my own thoughts: for me, Ilha de Guaratiba is "the place par excellence". The cradle of my history and geography, the place has always welcomed me throughout my years of existence: the place where my parents were born and where I was born and raised. The place where I studied and worked. The place where I now teach in the schools where I learned to love geography. The place where I met Jeane, my wife, and where Nicole, my daughter, took her first steps. The place that formed me and made me who I am today. A place where I had my deepest experiences. As far as I'm concerned, there's no place like Ilha de Guaratiba.

6.6 **Flowing into the Olympic River**

What does the drilling of the Grota Funda Tunnel mean for the urbanization march that is underway on Ilha de Guaratiba? As resident Càtia sees it,

> The Grota Funda Tunnel represents a door that has opened recently, a kind of portal that has brought good things and, in equal measure, a lot of insecurity. The good thing is that it has made life easier for

residents who, like me, used to lose important hours of their day crossing the mountains. I work in Botafogo and therefore suffered a lot in traffic. Both on the way to the office in the morning and on the way home during rush hour, traffic would come to a standstill on Serra da Grota Funda. Despite this positive side, after the opening of the tunnel I felt much more insecure than before. With the tunnel, people discovered our neighborhood, causing violence to increase here. Free and fast access to our place became a two-way street for us residents. The city that was on the other side of the mountains is now on our doorstep (Catia - executive secretary - 35 years old).

In the continuing sprawl of the city of Rio de Janeiro, landfills of lakes, ponds, mangroves, swamps and even significant advances over the sea contributed to the city's expansion (ABREU, 1987; 2008; GALVÀO, 1992). On the other hand, the opening of tunnels became necessary due, above all, to the physical characteristics of our city, gifted by nature with mountainous massifs and forest domains (Tijuca, Pedra Branca and Gericinó-Mendanha). Therefore, it is worth repeating that, since 1887, the drilling of tunnels became a viable solution to connect the different parts of the city separated by elevations that needed to be broken (ABREU, 1987; 2008; GALVÀO, 1992; MELLO, 1995; CARVALHO, 2002). As this process continued, in the words of resident Càtia, "the Grota Funda Tunnel represents an open door between Ilha de Guaratiba and the city that was on the other side of the mountain".

What does the Tûnel da Grota Funda represent for a traditional farmer used to the isolation of the place and, disconcertingly, its country town characteristics? For an answer, let's take a look at what Mr. Agostinho says:

Since the tunnel was pierced, we've seen nothing but growth. The place has changed a lot since the tunnel. A while ago, I used to look down and see nothing but plantations. Today you can see the number of farms, houses and condominiums. After the tunnel, even more people moved in. What's worse, they've only been rude. The other day I was down there and all I could see was the number of people walking past. All strange people who don't have the courage to say good morning (Mr. Agostinho - farmer and beekeeper - 71 years old).

Mr. Agostinho lives in a true paradise, literally in the middle of the forest, on one of the hills that surround the place. Like most of the old farmers in the area, Agostinho is the son of Portuguese. His parents arrived in Ilha de Guaratiba before he was born, in the 1920s, settling there and preserving it as their home and source of livelihood through their banana plantation and beekeeping. The resident in question understands that the Grota Funda Tunnel, as well as having transformed the place with the growing number of houses, has changed the interpersonal relationships he was used to. For him, a simple greeting is part of everyday life and the necessary interaction between his peers. The friendly farmer, who loves to chat, is indignant at the indifference with which he is treated by the new residents who, because they have a different attitude towards each other, are not given to the camaraderie and neighborly relations with which Mr. Agostinho is accustomed. At the confluence of time, he adopts an attitude and even has a feeling of strangeness towards those who are arriving on Guaratiba Island. The experience of this long-time Guaratibano is an illustration of one of the changes in the life of relationships caused both by the confluence of Rio de Janeiro's urbanization on the place, and by the flow of Ilha de Guaratiba into the waters of an Olympic River that flows with everyone.

FINAL WORDS

This book is the result of a long academic career. This journey highlights the dynamism in which the place being researched is involved. In fact, approaches to the multidimensionality inherent in the lived universe of Guaratiba have accompanied me since the dawn of my academic journey. On this path, the dimension of what is lived has contributed to elevating the neighborhood that Guaratibanos love to the status of a place, an approach that was initially written as an undergraduate monograph in geography at the Moacyr Sreder Bastos University Center in Campo Grande, located in the West Zone of Rio de Janeiro (FERNANDES, 2003). This research was further developed and gave rise to a new work, in which we highlighted the role of the (land/real estate) valuation of space in the process of place valuation, defended as such in the course on Territorial Policies in the State of Rio de Janeiro, at the level of Specialization, at UERJ (FERNANDES, 2006). Three years later, I obtained my master's degree by defending my master's thesis, fulfilling the requirements of the course at UERJ's Post-Graduate Program in Geography. To do this, I looked at the individual and collective geographies of the informal geographers of Ilha de Guaratiba. On this journey, I sought to base myself on the precepts of the exponents of humanistic geography Lowenthal and Tuan, understanding human beings as capable of talking about their lived worlds (FERNANDES, 2010).

The unfolding of my concerns about Ilha de Guaratiba has captivated me to continue writing about the place. Furthermore, to paraphrase Tuan (1961, p. 32), "I have no obligation to describe any area other than that for which I have a special affection or an inexplicable fascination". The place in which I live is and always will be my fundamental point of concern.

Throughout this text, we have seen that various scientists - especially since the 1970s - have been committed to criticizing the reductionist view of man postulated by positive science, a trend that has helped humanistic geographers to interpret feelings and understand the relationship between men and their world. This current of geographical thought is also opposed to positivism, in an attempt to overcome its reductionism. This thesis is concerned precisely with this perspective, which is to translate/decode the soul of a particular place, namely Ilha de Guaratiba.

By decoding the lived experiences of members of the community studied, this thesis has underlined the phenomenological premise that "each individual constructs their own world" (SCHUTZ, 1979, p.17). More than that, by deciphering experiences, feelings and other experiences common to individuals, it revealed the intersubjective phenomenon, since "individual experiences converge into the collective" (MELLO, 2000, p.42). Through the niche of shared experiences, including those of the researcher, I was able to unveil the memorable geographies of Ilha de Guaratiba, as well as the reflection of the transformations taking place in the lives and in the set of values and feelings of the Guaratibanos, who are also involved in this spatial and existential metamorphosis taking place in their lived universe.

In order not to end this work amid methodological concerns and justifications, I would like to evoke the sacred nature of the sense of place, drawing on Mircea Eliade's (2007) deep-rooted religious perspective in relation to the mythical description of the symbolism of the center. Alluding to this sacred approach represents a final attempt to express - through words - the value, meaning, importance and relevance of the place for the person who experiences it and makes it their central axis towards which everything converges. The aforementioned philosopher, when studying the history of religions, goes through some Eastern traditions until he places Paradise, where man (Adam) was created by God from clay, at the center of the world - or at the navel of the Earth (ELIADE, 2007; 2012). Eliade's insight takes us back to a kind of reverie in an attempt to ascertain why human beings love and revere their lived universe.

What was the greatest punishment imposed on man for his disobedience to the Divine order? Some would say it was the exhaustion imposed by work, which from then on became tiring. The majority would certainly say that death, which came into force after sin, was the "capital punishment" imposed on man. Physical death, however, was not imposed by the Creator, but came upon man as the greatest consequence of his sin. Once death was ruled out, wasn't expulsion from Paradise - the place par excellence - the great punishment imposed on man by the Creator? Amid the march of urbanization and the sprouting of new memorable geographies that characterize the "race to the west of Rio", what is in store for Ilha de Guaratiba and its residents?

In fact, in the course of an Olympic Rio de todos gente and with the formation of new meanders, the Grota Funda Tunnel and the road network, leveraged by the construction of the Transoeste highway, appear to meet the desires of residents - and even frequenters of Ilha de Guaratiba - and the ambition of the public authorities to incorporate this area into the city.

Ilha de Guaratiba, inaugurated as a sesmaria in 1579, has thrived for centuries thanks to its agriculture and the peaceful life of its inhabitants. In recent years, however, and with the prospect of the Olympics, Rio's waterway has grown in two movements: from Guaratiba to Rio proper, and from Rio, at the confluence of time, towards the land of Mr. Willian, the landscaper Roberto Burle Marx and the Guaratibanos who are entwined with their lived universe.

REFERENCES

ABBAGNANO, Nicola. *Dictionary of Philosophy.* Sao Paulo: Martins Fontes, 2007. 1210 p.

ABREU, Mauricio de Almeida. *Urban evolution of Rio de Janeiro.* Rio de Janeiro: IPLANRIO, 1987. 147 p.

. The City, the Mountain and the Forest. In: ABREU, Mauricio de Almeida. *Nature and Society in Rio de Janeiro.* Rio de Janeiro: Biblioteca Carioca, 1992. p. 55-103.

. _The Appropriation of Territory in Colonial Brazil. In: CASTRO, Ina Elias de; GOMES, Paulo César da Costa; CORRÊA, Roberto Lobato (Org). *Geographical Explorations:* Journeys at the End of the Century. Rio de Janeiro: Bertrand Brasil, 1997. p.197-245.

. On the Memory of Cities. *Revista da Faculdade de Letras* - Geografia I série, Porto, v.14, p. 77-97, 1998.

. Cities: Spatialities and Temporalities. In: CARLOS, Ana Fani Alessandri; LEMOS, Amàlia Inês Geraiges (Orgs). *Urban Dilemmas:* new approaches to the city. Sao Paulo: Contexto, 2003. 430 p.

. *Rio de Janeiro's urban evolution.* Rio de Janeiro: IPP, 2008. 155 p.

ABREU, Regina. Do I mix chewing gum with bananas? On the relationship between theory and research in social memory. In: GONDAR, Jô; DODEBEI, Vera (Orgs). *What is social memory*? Rio de Janeiro: Contra Capa Livraria, 2011. 160 p.

ALMEIDA, Maria Geralda de. The songs and enchantments of a backlands geography by Patativa do Assaré. In: MARANDOLA JR, Eduardo; GRATAO, Lùcia Helena Batista (Orgs). *Geography and*

Literature: essays on geography, poetry and imagination. Londrina: Eduel, 2010. 354 p.

ANTONELLO, Ideni Terezinha. Amazonian territorialities shine through in Peregrino Jùnior's literary narrative. In: MARANDOLA JR, Eduardo; GRATÀO,

Lùcia Helena Batista (Orgs). *Geography and Literature*: essays on geography, poetics and imagination. Londrina: Eduel, 2010. 354 p.

ARANHA, Maria Lùcia de Arruda. *History of Education*. São Paulo: Moderna, 1996. 255 p.

ARANHA, Maria Lùcia de Arruda; MARTINS, Maria Helena Pires. *Philosophy themes.* Sâo Paulo: Editora Moderna, 1992. 232 p.

ARAÙJO, Dorothy Sue Dunn de. The vegetation of the Guaratiba-Sepetiba lowlands. In: KNEIP, Maria Lina et al. *Prehistoric Collectors and Fishermen of Guaratiba-Rio de Janeiro.* Rio de Janeiro: UFRJ; Niterói: EDUFF, 1987. p. 47-72.

ASSIS, Lenilton Francisco de. Turismo de Segunda Residência: a Expressão Espacial do Fenômeno e as Possibilidades de Análisis Geogràfica. *Revista Territòrio,* Rio de Janeiro: Sep/Oct, p. 107-122, 2003.

ATLAS of nature conservation units in the state of Rio de Janeiro, 1990. Irregular pagination.

BACHELARD, Gaston. *The poetics of space*. Sâo Paulo: Martins Fontes, 2008. 242 p.

BANDEIRA, Lucia Batista. Mallet - *A Neighborhood Elected and Demarcated Affectively*. 1999. Monograph (Specialization in Territorial Policies in the State of Rio de Janeiro) - Institute of Geography, State University of Rio de Janeiro, Rio de Janeiro, 1999.

BASTOS, Ana Regina Vasconcelos Ribeiro. Space and Literature: some theoretical reflections. *Revista Espaço e Cultura,* Rio de Janeiro, n.5, 1998.

BELL, Daniel. *The Advent of Post-Industrial Society*. Sao Paulo: Ed. Cultrix, 1977.

BERDOULAY, Vicent; ENTRIKIN, John Nicholas. Place and Subject: theoretical perspectives. In: MARANDOLA JR, Eduardo et al. *Qual o espaço do lugar?*: Geografia, epistemologia, fenomenologia. Sao Paulo: Perspectiva, 2012. 307 p.

BERTA, Ruben. Guaratiba: Urban Structuring Plan foresees 4-storey buildings. *O Globo newspaper*, Rio de Janeiro, April 28, 2012.

BONNEMAISON, Joel. Journey around the Territory. In: CORRÊA, Roberto Lobato; ROSENDAHL, Zeny (Org). Cultural Geography: a Century (3). Rio de Janeiro: EdUERJ, 2002. p. 83-131.

BORBA, Francisco da Silva. *UNESP dictionary of contemporary Portuguese*. Curitiba: Pia, 2011. 1488 p.

BOSSÉ. Mathias Le. Questions of identity in cultural geography - Some contemporary conceptions. In: CORRÊA, Roberto Lobato; ROSENDAHL, Zeny. *Cultural Geography* - An Anthology. V. 2. Rio de Janeiro: EdUERJ, 2013. 292 p.

BOSI, Ecléa. *Memória & sociedade* - Lembrança de velhos. Companhia das Letras: Sao Paulo, 2003. 488 p.

BUTTERFIELD, Roger. "Henry Ford, the Wayside Inn, and the Problem of 'History is Bunk'". *Massachusetts Historical Society Preccedings*, Massachusetts, v. 77, p. 53-66,1965.

BUTTIMER, Anne. Apprehending the Dynamism of the Lived World. In: CHRISTOFOLETTI, Antônio (Org). *Perspectives on Geography*. Sao Paulo: DIFEL, 1982.

p. 165-193.

BUTTIMER, Anne. Home, Horizon of Reach and the Sense of Place. *Revista Geograficidade*, Niterói, v.5, n.1, p. 4-19, verâo 2015.

CALS, Soraia. *Roberto Burle Marx*: a *Photobiography*. Rio de Janeiro: Bolsa de Arte, 1995. Irregular pagination.

CANCLINI, N. G. *La Globalización Imaginada*. Buenos Aires: Ed. Paidós, 1999.

CAPEL, Horâcio. *The Cosmopolis and the City.* Barcelona: Ed. Del Serbal, 2003.

CARLOS, Ana Fani Alessandri. *Place in the World*. Sâo Paulo: Hucitec, 1996.

CARVALHO, Ronaldo Cerqueira de. *Rio de Janeiro:* A city connected by tunnels

- a panorama up to the end of the 1960s. 2002. Monograph (Specialization in Geography) - Institute of Geography, Rio de Janeiro State University, Rio de Janeiro, 2002.

CARVALHO, Ronaldo Cerqueira de. Rio de Janeiro - a city connected by tunnels. Rio de Janeiro: Secretaria Municipal de Urbanismo; Instituto Municipal de Urbanismo Pereira Passos, 2004. 57 p.

CASSIRER, Ernest. *The Philosophy of Symbolic Forms*: II - mythical thought. Sâo Paulo: Martins Fontes, 2004. 432 p.

. *The Philosophy of Symbolic Forms*: III - phenomenology of knowledge. Sâo Paulo: Martins Fontes, 2011. 818 p.

CASTELLS, Manuel. *The Network Society*. Sâo Paulo: Paz e Terra, 2002.

CASTRO, Augusto César de. *Guaratiba*: Yesterday and Today. Course Conclusion Paper (History Degree) - History Degree Course, Fundaçâo Educacional Unificada Campograndense (FEUC), Rio de Janeiro, 2002.

CHRISTOFOLETTI, Antônio. The Perspectives of Geographical Studies. In: CHRISTOFOLETTI, Antônio. *Perspectives on Geography*. Sâo Paulo: DIFEL, 1982. p. 11-36.

CLAVAL, Paul. The Role of the New Cultural Geography in Understanding Human Action. In: CORRÊA, Roberto Lobato; ROSENDAHL, Zeny (Org). *Matrices of Cultural Geography.* Rio de Janeiro: EdUERJ, 2001. p. 35-86.

. *Cultural Geography*. Florianópolis: Editora da UFSC, 2007. 453 p.

CORRÊA, Magalhâes. O Sertâo Carioca. *Revista do Instituto Histórico e Geogràfico Brasileiro,* Rio de Janeiro, v. 167, 478p., 1936.

CORRÊA, Roberto Lobato. The Environment and the Metropolis. In: ABREU, Mauricio de Almeida. *Nature and Society in Rio de Janeiro*. Rio de Janeiro: Biblioteca Carioca, 1992. p. 27-36.

. *Urban Space*. 4.ed. Sâo Paulo: Atica, 2000. 94 p.

CORRÊA, Roberto Lobato. Space: A Key Concept in Geography. In: CASTRO, Ina Elias de; GOMES, Paulo César da Costa; CORRÊA, Roberto Lobato (Org).

Geography: concepts and themes. 4.ed. Rio de Janeiro: Bertrand Brasil, 2002. p. 15-47.

. Cultural Geography and the Urban. In: CORRÊA, Roberto Lobato; ROSENDAHL, Zeny (Org). *Introduction to cultural geography*. Rio de Janeiro: Bertrand Brasil, 2003. p. 167-186.

. On cultural geography. In: Textos NEPEC (volume 3). Rio de Janeiro: UERJ, 2007.

. The spatiality of culture. In: CD cultura. Rio de Janeiro: UERJ, 2011.

CORRÊA, Roberto Lobato; ROSENDAHL, Zeny. Cultural Geography: presenting an anthology. In: CORRÊA, Roberto Lobato; ROSENDAHL, Zeny (Orgs). *Cultural Geography:* an anthology (1). Rio de Janeiro: EdUERJ, 2012. 344 p.

COSGROVE, Denis. Towards a radical cultural geography: problems of theory. In: CORRÊA, Roberto Lobato; ROSENDAHL, Zeny (Org). *Introduction to cultural geography*. Rio de Janeiro: Bertrand Brasil, 2003. 224 p.

. Geography Is Everywhere: Culture and Symbolism in Human Landscapes. In: CORRÊA, Roberto Lobato; ROSENDAHL, Zeny (Org). *Landscape, Time and Culture.* Rio de Janeiro: EdUERJ, 2004. p. 92-123.

COSTA, Mariana Timóteo da. Two steps away from paradise. *O Globo Magazine*, Rio de Janeiro, December 15, 2013.

COSTA, Otàvio. Memory and Landscape: in search of the symbolism of places. *Revista Espaço e Cultura*, Rio de Janeiro, Commemorative edition 1993-2008, p. 149-156, 2008.

DARDEL, Eric. *Man and the Earth*: the nature of geographical reality. Sâo Paulo: Perspectiva, 2011. 159 p.

DIAS, Alice Ferreira Rodrigues. Guaratiba Island: Whose Green Landscape? *Espaço Aberto Magazine*, Rio de Janeiro, v.1, n.1, p. 89108, 2011.

DREW, David. *Human-Environment Interactive Processes*. Rio de Janeiro: Bertrand Brasil, 2002. 224 p.

ELIADE, Mircea. *The Nostalgia of Origins*. Paris: Folio-Essais; Galimard, 1971.

Myth of the eternal return. Sâo Paulo: Mercuryo, 2007. 175 p.

ELIADE, Mircea. *Images and symbols*: an essay on mystical-religious symbolism. Sâo Paulo: Martins Fontes, 2012. 178 p.

ELIAS, Norbert & SCOTSON, John. *The Established and the Outsiders*: Sociology of Power Relations in a Small Community. Rio de Janeiro: Jorge Zahar, 2000 224 p.

ENTRIKIN, John Nicholas. Contemporary Humanism in Geography. *Bulletin of Theoretical Geography*, São Paulo, v. 10, 1980.

FARAH, Fàbio. José Humberto Resende - The doctor from Sudàrio. *Isto é Gente magazine*, Sâo Paulo, June 2, 2003. Available at: <http: //www.terra.com.br/istoegente/ 200/reportagens/jose_humberto_resende_sudario.htm>. Accessed on: November 20, 2015.

FERNANDES, Marcio Luis. *Guaratiba Island*: From Space to Place. 2003. 44f. Monograph (Graduation in Geography) - Moacyr Sreder Bastos University Center, Rio de Janeiro, 2003.

. *Guaratiba Island* in its natural attributes. Rio de Janeiro, 2005. 41f. Available at: <http://www.webartigos.com/artigos/ilha-de- guaratiba-em-seus-atributos-naturais/136345/>. Accessed on: October 19, 2015.

. *The valorization of "Space" producing the valorization of "Place*:" The case of Ilha de Guaratiba - R.J. 2006. 56 f. Monograph (Specialization in Geography) - Institute of Geography, Rio de Janeiro State University, Rio de Janeiro, 2006.

. *Guaratiba Island and its landscapes*. Rio de Janeiro, 2008. 32 f. Available at: <http://www.webartigos.com/artigos/ilha-de- guaratiba-e-suas-paisagens/ 136627/>. Accessed on: October 19, 2015.

. For a Necessary Change in Values: a proposal for the production of (urban) space that favors use and not exchange. In: SIMPÓSIO NACIONAL O RURAL E O URBANO NO BRASIL, 2., 2009, Rio de Janeiro. *Proceedings*...Rio de

Janeiro: UERJ, 2009.

FERNANDES, Marcio Luis. *Decoding past and present geographies of Ilha de Guaratiba.* 2010. 99f. Dissertation (Master's in Geography) - Institute of Geography, Rio de Janeiro State University, Rio de Janeiro, 2010.

. *Descortinando a marcha urbanizadora em Ilha de Guaratiba a partir das experiências vividas pelos seus moradores.* Rio de Janeiro, 2011. 25 f. Available at: <http://www.webartigos.om/artigos/descortinando-a-marcha- urbanizadora - em-ilha-de-guaratiba-partir-das-experiencias-vidas-por-seusmores/136505/>. Accessed on October 19, 2015.

. The Identity Character of Toponymy. In: CONGRESSO INTERNACIONAL DO NÙCLEO DE ESTUDO DAS AMÉRICAS, 3., 2012. Rio de Janeiro. *Proceedings...* Rio de Janeiro: UERJ, 2012.

. Globalization and urbanization of the world. Rio de Janeiro, 2013. Available at: <http://www.webartigos.com/artigos/globalizacao-e- urbanizacao-do-mundo/136480/>. Accessed on October 19, 2015.

. _Another horizon in search of the humanization of geography. *Revista Geograficidade*, Niterói, v.4, n.1, p. 78-87, verâo 2014.

. _Race to the West Zone. *O Dia*, Rio de Janeiro, November 26, 2014 b. Opinion column.

. _Unveiling the symbolic universe of a place. Revista Perspectiva Geogràfica, Cascavel, v.9, n.11, 2015.

FERREIRA, Antônia Maria M; oLIVEIRA, Marli Vieira de. Contribution to the Archaeo-Geological Study of the Upper Quaternary of the Baixada de Guaratiba-Sepetiba. In: Kneip, Maria Lina et al. *Prehistoric Collectors and Fishermen of Guaratiba-Rio de Janeiro*. Rio de Janeiro: UFRJ; Niterói: EDUFF, 1987. p. 29-45.

FIREY, Walter. Sentiments and symbolism as ecological variables. *American Sociological Review*, v.10, n.2, p. 140-148, 1945 (Annual Meeting Papers).

FIREY, Walter. Feelings and symbolism as ecological variables. In: ROSENDAHL, Zeny; CORRÊA, Roberto Lobato. *Cultural geography*: an anthology, volume II. Rio de Janeiro: EdUERJ; 2013. p. 21-34.

FREITAS, Inês Aguiar de; PERES, Waldir Rugero; RAHY, Ione Salomao. Hitler's Window. *GeoUERJ - Journal of the Geography Department*. Rio de Janeiro n. 6, p. 29-36, 1999.

FRÉMONT, Armand. Histoire d'une Recherche. In: Frémont et al. *Living Space and Civilizations.* Paris: CNRS, 1982. 106 p.

FRIDMAN, Fania. *Donos do Rio em nome do rei*: uma história fundiària da cidade do Rio de Janeiro. Rio de Janeiro: Jorge Zahar Editor (Editora Garamond), 1999. 304 p.

GALLAIS, Jean. Some Aspects of Lived Space in the Civilizations of the Tropical World. In: CORRÊA, Roberto Lobato; ROSENDAHL, Zeny (Org). *Cultural Geography*: a century (3). Rio de Janeiro: EdUERJ, p. 63-81, 2002.

GALVÀO, Maria do Carmo Corrêa. Focus on the environmental question in Rio de Janeiro. In: ABREU, Mauricio de Almeida. *Nature and Society in Rio de Janeiro*. Rio de Janeiro: Biblioteca Carioca, 1992. p. 13-26.

GEIGER, Pedro Pinchas. Contribution to the debate on urban spatialities and temporalities. In: CARLOS, Ana Fani Alessandri; LEMOS, Amàlia Inês Geraiges (Orgs). *Urban Dilemmas*: new approaches to the city. Sâo Paulo: Contexto, 2003. 430 p.

GEERTZ, Clifford. *The interpretation of cultures*. Rio de Janeiro: LTC, 2013. 213 p.

GOMES, Paulo César da Costa. *Geography and Modernity*. Rio de Janeiro: Bertrand Brasil, 2007. 366 p.

GONDAR, Jô; DODEBEI, Vera (Orgs). *What is social memory?* Rio de Janeiro: Contra Capa Livraria, 2011. 160 p.

GUERRA, Antônio Teixeira. *Geological and Geomorphological Dictionary. 8.* ed. Rio de Janeiro: IBGE, 1993. 446 p.

HAESBAERT, Rogério. *O Mito da Desterritorialização*: Do "fim dos Territórios" à Multiterritorialidade. Rio de Janeiro: Bertrand Brasil, 2004. 400 p.

. _From deterritorialization to multiterritoriality. In: ENCONTRO DE GEÓGRAFOS DA AMÉRICA LATINA - EGAL, 10., 2005, Sâo Paulo. *Proceedings...* Sâo Paulo: USP, 2005.

HALL, Stuart. *Representations:* cultural representations and signifying practices. London: Routledge publications, 1997.

HALBWACHS, Maurice. *Collective Memory*. Presses Universitaires de France: Paris, 1968.

. *Collective Memory*. São Paulo: Centauro, 2013. 224 p.

HALLEY, Bruno Maia. The Neighborhood and the plots of place. Revista Geograficidade, Niterói, v.4, n.1, p. 43 - 57, Verâo 2014.

HERDY, José Maria. Ilha de Guaratiba has everything it takes to become an efficient neighborhood.In:: PORTAL Guaratiba, Rio de Janeiro, 01 jun. 2013. Available at:<http://www.portalguaratiba .com.br/2013/noticias/010601_ilha_de_guaratiba_tem_tudo_para_se_t ornar_um_bairro_eficiente.html>. Accessed on: November 20, 2015.

HARVEY, David. *The Postmodern Condition*. São Paulo: Ediçôes Loyola, 1989.

HARVEY, David. *The Postmodern Condition*: an inquiry into the origins of cultural change. Sâo Paulo: Ediçôes Loyola, 1992. 349 p.

. *The Capitalist Production of Space*. Sâo Paulo: Annablume, 2005.

HIGHET, Gilbert. *The Classical Tradition*: Greek and Roman Influences on Western Literature. Oxford: Clarendon, 1949.

HOLZER, Werther. *Humanist geography:* its trajectory from 1950 to 1990 (Master's dissertation). Rio de Janeiro: PPGEO/UFRJ, 1992. 550 f.

. Eric Dardel's Phenomenological Geography. In: CORRÊA, Roberto Lobato; ROSENDAHL, Zeny (Org). *Matrices of cultural geography*. Rio de Janeiro: EdUERJ, 2001. p. 103-122.

. Humanist Geography: a review. *Revista Espaço e Cultura*, Rio de Janeiro,

Commemorative edition 1993-2008, p. 137-147, 2008.

JANOT, Luiz Fernando. On the road to Guaratiba. *O Globo newspaper*, Rio de Janeiro, October 26, 2013. Opinion column.

JAPIASSÙ, Hilton; MARCONDES, Danilo. *Didonărio bàsico de filosofia*. Rio de Janeiro: Zahar, 2006. 309 p.

JOHNSTON, Ronald John. Introduction to the international study of the history of geography. In: JOHNSTON, Ronald John; CLAVAL, Paul (Orgs). *Current geography: geographers and trends.* Barcelona: Ariel, 1986 a. p. 13-25.

JOHNSTON, Ronald John. *Geography and Geographers*: Anglo-American human geography since 1945. Sâo Paulo: Difel, 1986. 359 p.

KNEIP, Maria Lina et AL (Org). *Prehistoric Gatherers and Fishermen of Guaratiba-Rio de Janeiro*. Rio de Janeiro: UFRJ; Niterói: EDUFF, 1987. 257 p.

KORYTOWSKI, Ivo. Literature and Rio de Janeiro. Rio de Janeiro, [20--]. Available at: <http://literaturaeriodejaneiro.blogspot.com.br/>. Accessed on: Nov. 20, 2015.

LAGO, Màrio. *Na Rolança do Tempo*. José Olympio: Rio de Janeiro, 1976. 350 p.

LEFEBVRE, Henri. *Presence and Absence*. Contribution to the theory of representations. Mexico: FCE, 1983. Irregular pagination.

. *The Urban Revolution*. Belo Horizonte: Editora UFMG, 2008.

LESSA, Carlos. *O Rio de Todos os Brasis*: Uma Reflexào em Busca de Auto-Estima. 2. ed. Rio de Janeiro: Record, 2001. 478 p.

LOWENTHAL, David. Geography, Experience and Imagination: Towards a Geographical Epistemology. In: CHRITOFOLETTI, Antônio. *Perspectives on Geography*. Sào Paulo: DIFEL, 1982. p. 103141.

. *The Past Is a Foreign Country*. Cambridge: Cambridge University Press, 1985.

. How we know the past. *Projeto História - Revista do programa de estudos pós-graduados de história*, Sào Paulo, v. 17, p. 63-201, 1998.

MARANDOLA JR, Eduardo. On Ontologies. In: MARANDOLA JR, Eduardo et al. *What is the space of place?* Geography, epistemology, phenomenology. Sào Paulo: Perspectiva, 2012. 307 p.

MASCARENHAS, Gilmar. *O Lugar da Feira-Livre na grande Cidade Capitalista: Conflito, Mudança e Persistência (Rio de Janeiro: 19641989).* 1991. 220 f. Dissertation (Master's Degree in Geography) - Federal University of Rio de Janeiro, Rio de Janeiro, 1991.

MASSEY, Doren. *Through Space*: A New Politics of Spatiality. Rio de Janeiro. Bertrand Brasil, 2008. 312 p.

MELLO, Joao Baptista Ferreira de. Geografia Humanistica: A Perspectiva da Experiência Vivida e Uma Critica Radical ao Positivismo. *Revista Brasileira de Geografia*, Rio de Janeiro, p. 91-115, 1990.

. *The Rio de Janeiro of Brazilian Popular Music Composers* - 1928/1991 - an introduction to humanistic geography. 1991. 300 f. Dissertation (Master's Degree in Geography) - Federal University of Rio de Janeiro. Rio de Janeiro, 1991.

. _The Humanization of Nature - an odyssey for the (re)conquest of paradise. In: SILVA, S. T.; VIANA, O. M. *Geografia e Questao Ambiental.* Rio de Janeiro: IBGE, 31-40, 1993.

. In Defense of Individuals in Geographical Studies. In: ENCONTRO NACIONAL DE HISTÓRIA DO PENSAMENTO GEOGRAFICO 1. 1999, Rio Claro. Proceedings...Rio Claro: UNESP, 1999. p. 113-118.

. *From Spaces of Darkness to Places of Extreme Luminosity* - The Universe of the Star Marlene as a Document for the Construction of Geographical Concepts. 2000. Thesis (Doctorate in Geography) - Federal University of Rio de Janeiro, Rio de Janeiro, 2000.

. Unveiling and (Re)thinking Spatial Categories Based on the Work of Yi-Fu Tuan. In: CORRÊA, Roberto Lobato; ROSENDAHL, Zeny (Org). *Matrices of cultural geography.* Rio de Janeiro: EdUERJ, 2001. p. 87-101.

. _The Restoration of Places of the Past. GeoUERJ Magazine, Rio de

Janeiro, n.12. p. 63-69, 2002.

. Symbols of Places, Spaces and "Displaces". *Space and Culture*. Rio de Janeiro, v. 16, p. 64-72, 2003.

MELLO, Joao Baptista Ferreira de. On the Pulse of the Marvelous City of Sao Sebastiao do Rio de Janeiro. In: SOCIEDAD LATINOAMERICANA DE ESTUDIOS SOBRE AMERICA LATINA Y EL CARIBE - SOLAR, 9, 2004, Rio de Janeiro. *Proceedings...* Rio de Janeiro: [S.n.], 2004.

. Values in Geography and the Dynamism of the Lived World in the Work of Anne Buttimer. *Space and Culture*. Rio de Janeiro, v. 19-20, p. 33-40, 2005.

. The Drums and Arrows of Sao Sebastiao do Rio de Janeiro. *Revista Imaginàrio e Arte,* Sao Paulo, n.15 p. 37-67, 2007.

. The River of Official and Vernacular Symbols. In: CORRÊA, Roberto Lobato; ROSENDAHL, Zeny (Org). *Space and culture*: thematic plurality. Rio de Janeiro: EdUERJ, 2008. p. 173-186.

MENDILOW, Adam Abraham. *Time and the novel*. London: John Spencer/Badger, 1960.

MENEZES, Luiz Fernando et al (Org). *Natural History of Marambaia*. Rio de Janeiro: EDUR, 2005.

MENEZES, Luiz Fernando; ARAÙJO, Dorothy Sue Dunn de; GOES, Maria Hilde de Barros. Marambaia: the last preserved restinga in Rio de Janeiro. *Revista Ciência Hoje*, v. 23, n.136, p. 28-37, mar. 1988.

MONBEIG, Pierre. *New studies in Brazilian human geography*. Sao Paulo: European Book Diffusion, 1957.

MORAES, Antônio Carlos Robert de. *Geografia:* pequena história critica. Sao Paulo: Annablume, 2007. 152 p.

MOREIRA, Ruy. *O pensamento geogràfico brasileiro*: as matrizes da renovaçao. Sao Paulo: Contexto, 2009. 172 p.

NOGUÉ Y FONT, J. The existential landscape of five groups of environmental experience. Methodological essay. In: BALLESTEROS, A. *Geografia y*

Humanismo. Barcelona: Oikos-tau, 1992. p. 87-96.

NORA, Pierre. *Between memory and history:* the problem of places. Sao Paulo: PUC-SP. 1993. p. 7-28. (Projeto História, n. 10).

OLIVEIRA, Livia de. The sense of place. In: MARANDOLA JR, Eduardo et al. *What is the space of place?* Geography, epistemology, phenomenology. Sao Paulo: Perspectiva, 2012. 307 p.

PALMER, Richard E. *Hermeneutics*. Sao Paulo: Martins Fontes, 1970. 284 p. (Ediçôes 70).

PANOFSKY, Erwin. *Meaning of the visual arts*. Sao Paulo: Perspectiva, 2004.

PARK, Robert Ezra. The City: suggestions for the investigation of human behavior in the urban environment. In: VELHO, Otàvio Guilherme. *The Urban Phenomenon*. Rio de Janeiro: Zahar, 1976. p. 2667.

PICKLES, Jonh. *Geography and humanism*. Norwich: Geo Books, 1985. 64 p.

PINTO, Rivadàvia. Guaratiba: A Pride of 407 Years. *Reason: the positive newspaper*. Rio de Janeiro, unpaginated, Nov. 1986.

POCOCK, Douglas. Place and the novelist. *Transactions of the Institute of British Geographers*: New Series, v. 6, p. 87-98, 1981.

PORTAL Guaratiba. Rio de Janeiro, [20--]. Available at: <http://www.portalguaratiba.com.br>. Accessed on: November 20, 2015.

DIGITAL PROCESSING: Geotechnologies and free software. [S.I.], [20--]. Available at: <http://www.processamentodigital.com.br>. Accessed on: November 20, 2015.

REDONDO, Andrea Albuquerque. The city grows towards Guaratiba. *Urbe Carioca.* Rio de Janeiro, September 25, 2012. Available at:< http:// urbe carioca.blogspot.com.br>.

Accessed on: Nov. 20, 2015.

RELPH, Edward. An inquiry into the relations between phenomenology and geography. *Canadian Geographer*, v. 14, n.3, p. 193-201, 1970.

. *Place and Placelessness*. London: Pion, 1976. 156 p.

. Reflections on the emergence, aspects and essences of place. In: MARANDOLA JR, Eduardo et al. *What is the space of place?* Geography, epistemology, phenomenology. São Paulo: Perspectiva, 2012. 307 p

RHEINGANTZ, Carlos. *First Families of Rio de Janeiro (16th and 17th centuries).* V. 1. Livraria Brasileira Editora: Rio de Janeiro, 1965. 556 p.

RIBEIRO, Luiz César de Queiroz. *Dos cortiços aos condominios fechados*: as formas de produçao da moradia na cidade do Rio de Janeiro. Rio de Janeiro: Civilizaçao Brasileira: IPPUR, UFRJ: FASE, 1997. 352 p.

RIBEIRO, Miguel Ângelo; COELHO, Maria do Socorro Alves. The importance of the second-home phenomenon and its implications for leisure-summer activities: the example of the state of Rio de Janeiro. In: ENCONTRO NACIONAL DA ANPEGE, 9., 2007, Niterói. *Proceedings...* Niterói: ANPEGE, 2007. 1 CD-ROM.

RIO 2016. Rio de Janeiro, [20--]. Available at: <http://www.rio2016.com>. Accessed on: Nov. 20, 2015.

ROSENDAHL, Zeny. *Space and Religion*: A Geographical Approach. Rio de Janeiro: EdUERJ: 2002. 90 p.

ROSENDAHL, Zeny. Space, Culture and Religion: Dimensions of Analysis. In: CORRÊA, Roberto Lobato; ROSENDAHL, Zeny (Org).

Introduction to cultural geography. Rio de Janeiro: Bertrand Brasil, 2003. 224 p.

ROTEIROS Geogràficos do Rio: free routes in defense of the city of Rio de Janeiro. Rio de Janeiro, [20--]. Available at: <http://www.roteirosdorio.com>. Accessed on: November 20, 2015.

RUA, Joâo. Urbanities and New Ruralities in the State of Rio de Janeiro: Some theoretical considerations. In: MARAFON, Glàucio José; RIBEIRO, Marta Foeppel (Org). *Studies in Fluminense Geography.* Rio de Janeiro: Infobook, 2002 A. p. 27-42.

. Urbanization in rural areas in the state of Rio de Janeiro In: MARAFON, Glàucio José; RIBEIRO, Marta Foeppel (Org). *Studies in Fluminense Geography.* Rio de Janeiro: Infobook, 2002 B. p. 43-69.

SA, Fàtima. Burle Marx Is Not Dead. *O Globo Magazine*, Rio de Janeiro, year 5, n. 227, November 30, 2008. Irregular pagination.

SACK, Robert David. Magic and space. *Annals of the association of American geographers*, v. 66, n.2, p. 309-322, 1976.

. *Human Territoriality*: its theory and history. Cambridge. Cambridge University Press, 1986. 256 p.

SALGUEIRO, Teresa Barata. Urban spatialities and temporalities. In: CARLOS, Ana Fani Alessandri; LEMOS, Amâlia Inês Geraiges (Orgs). *Urban Dilemmas*: new approaches to the city. São Paulo: Contexto, 2003. 430 p.

SANTOS, Milton. *Space and Method*. Sâo Paulo: Nobel, 1992. 88 p.

. *Metamorphoses of Inhabited Space*. 5. ed. Sâo Paulo: Hucitec, 1997. 117 p.

. *For another globalization*. Rio de Janeiro: Record, 2001.

SANTOS, Milton. *The Nature of Space:* Technique and Time, Reason and Emotion. Sao Paulo: EDUSP, 2002. 384 p.

SANTOS, Noronha. *The parishes of old Rio*. Rio de Janeiro: Ediçôes O Cruzeiro, 1965.

SCHUTZ, Alfred. *Phenomenology and social relations*. Rio de Janeiro: Zahar, 1979. 319 p.

SEAMON, DAVID. Body-subject, time-space routines, and placeballets. In: BUTTIMER, Anne and SEAMON, David. *The Human Experience of Space and Place*. New York: St. Martin's Press, 1980. p. 148-165.

SEAMON, DAVID. Body-Subject, Spatio-Temporal Routines and Dances of Place (translated by Paulo Mauricio Rangel). *Revista Geograficidade*, Niterói, v.3, n.2, p. 4-18, Winter 2013.

SILVA, Armando Corrêa da. *Space out of place*. Sâo Paulo: Hucitec, 1988. 128 p.

SILVA, Kelly Cristina Rodrigues. Memory for thinking about space: the perspective of place. *Revista Geograficidade*, Niteroi, v.5, n.2, p. 2637, Winter 2015.

SIMMEL, Georg. The Metropolis and Mental Life. In: VELHO, Otâvio Guilherme. *The Urban Phenomenon*. Rio de Janeiro: Zahar, 1976.

Roberto Burle Max SITIO. Rio de Janeiro, [20--]. Available at: <http://sitioburlemarx.blogspot.com.br>. Accessed on: November 20, 2015.

SOARES, Maria Therezinha de Segadas Soares. *Aspectos da geografia carioca*: Rio de Janeiro: Conselho Nacional de Geografia, 1962.

. The geographical concept of neighborhood and its exemplification in the city of Rio de Janeiro. In: BERNANDES, Lysia; SOARES, Maria

Therezinha de Segadas. *Rio de Janeiro - City and Region.* Rio de Janeiro: Biblioteca Carioca, 1990. p. 105-120.

SOUZA, Marcelo Lopes de. O bairro contemporâneo: ensaio de abordagem política. *Revista Brasileira de Geografia*, Rio de Janeiro, p. 139-172, 1989.

. *Changing the City*: A Critical Introduction to Urban Planning and Management. 3. ed. Rio de Janeiro: Bertrand Brasil, 2004. 556 p.

. *ABC do Desenvolvimento Urbano*. 2.ed. Rio de Janeiro: Bertrand Brasil, 2005. 190 p

TARGINO, Tânia; MONTEIRO, Neide Carvalho. *Atlas Escolar da Cidade de Rio de Janeiro.* 1. ed. Rio de Janeiro: Secretaria Municipal de Educaçâo: Instituto Municipal de Urbanismo Pereira Passos, 2000. 41p.

TUAN, Yu Fu. Topophilia or sudden encounter with landscape. *Landscape*, v.11, n. 1, p. 29-32, 1961.

. Sacred Space: Exploration of an Idea. In: DIMENSION of human geography. Chicago: University of Chicago, 1978. p. 84-100.

. *Topophilia*: A Study of Perception, Attitudes and Values of the Environment. Sâo Paulo/Rio de Janeiro: DIFEL, 1980. 288 p.

. Humanistic Geography. In: CHRITOFOLETTI, Antônio. *Perspectives on Geography*. Sâo Paulo: DIFEL, 1982. p. 143-164.

. *Space and Place*: The Perspective of Experience. Sâo Paulo: DIFEL, 1983. 250 p.

. *The good life*. Madison: The University of Wisconsin Press, 1986. 191 p.

. A view of geography. *Geographical Review*. New York, v. 81 n. 1: p. 99-106, 1991.

TUAN, Yu Fu. *Escapism*. Baltimore: The Johns Hopkins University Press, 1998. 245 p.

. *Landscapes of Fear.* Sâo Paulo: UNESP, 2005. 373 p.

. Space, Time and Place: A Humanist Framework. *Revista Geograficidade*, Niterói, v.1, n.1, p. 8-19, Winter 2011.

. *Topophilia:* A Study of Perception, Attitudes and Values of the Environment. Londrina: Eduel, 2012. 344 p.

. *Space and Place:* The Perspective of Experience: Londrina, PR: EDUEL, 2013. 248 p.

WEID, Elisabeth von der. *The streetcar as an element of urban expansion in Rio de Janeiro.* Rio de Janeiro: Casa de Rui Barbosa Foundation, 1997. 30 p.

WIRTH, Louis. Urbanism as a way of life. In: VELHO, Otàvio Guilherme. *The Urban Phenomenon.* Rio de Janeiro: Zahar, 1976. p. 90113.

WRIGHT, John. Terrae incognitae: the place of the imagination in geography. *Annals of the Association of American Geographers*, v. 37. p. 01-15, 1947.

YAZIGI, Eduardo. Urban Environmental Heritage: remaking a concept for urban planning. In: CARLOS, Ana Fani Alessandri; LEMOS, Amâlia Inês Geraiges (eds). *Urban Dilemmas*: New Approaches to the City. Sâo Paulo: Contexto, 2003. p. 253-265.

ANNEX

ANNEX A - Decree No. 3158 of July 23, 1981. Establishes the Denomination, Codification and Delimitation of Neighborhoods in the City of Rio de Janeiro (Summary Restricted to the Guaratiba Administrative Region).

THE MAYOR OF THE CITY OF RIO DE JANEIRO, in the use of his legal attributions, in view of Deliberation No. 114, of April 8, 1981 of the City Plan Commission - P/COPLAN, approved in case No. 02/1476/80,

DECREE:

Art. 1 - The denomination, codification and delimitation of the neighborhoods of the City of Rio de Janeiro are established by this decree.

Art. 2 - The neighborhoods are listed in Annex I with their name and codification.

Sole Paragraph - Neighborhoods are coded according to the following criteria:

- the first digit corresponds to the Planning Area number;
- the next two digits correspond to the number of the Administrative Region;
- the last three digits correspond to the neighborhood number.

Art. 3 - The delimitation of neighborhoods is established in Annex II.

Art. 4 - This decree shall enter into force on the date of its publication, revoking any provisions to the contrary.

Rio de Janeiro, July 23, 1981 - 417th anniversary of the founding of the city

JULIO COUTINHO

Joaquim Torres Araujo

Carlos Alberto de Carvalho

José Maria da Motta

Vicente de Paulo Barreto

Lucy Serrano Ribeiro Vereza

Paulo Cesar Catalano

Renato da Silva Almeida

ANNEX I: NAMING AND CODING OF NEIGHBORHOODS

XXVI Administrative Region

526151 Guaratiba

526152 Barra de Guaratiba

526153 Pedra de Guaratiba

ANNEX II: DELIMITATION OF NEIGHBORHOODS

XXVI RA - GUARATIBA

151 - **Guaratiba**

From Praia da Capela, on the Bay of Sepetiba, along the extension of Avenida Carlos da Silva Rocha and, on this (excluded) crossing Estrada da Matriz, to the junction with Rua Soldado Prim Canes; from this point, along the extension of the alignment of Avenida Carlos da Silva Rocha, to Rua 25 do PAL 18.529; along this (included) to Estrada da Capoeira Grande; along this (included) to Estrada do Catruz; along this (excluded) to Rua Maestro Deozilio; along this (excluded) and its extension, to Baia de Sepetiba; from there, along the coast, passing through Paia do Aterro (included) to the mouth of the Rio do Ponto or Piai, at Saco do Piai; along the riverbed, crossing Estrada do Piai, to Estrada da Pedra; along the latter (included) to the junction with Rua General Alexandre Barreto (excluded); from there, up the spur to the 157m elevation point on Serra do Cantagalo; from this point, along the ridge in a northeasterly direction, to the 203m elevation point; from this point, in a northerly direction, passing through elevation points 178m, 194m, 157m and 153m, to elevation point 249m on Morro de Santa Eugênia, from this point, along a straight line in an easterly direction, to elevation point 227m on Serra de Inhoaiba; from this point, along the ridge and the spur, passing through elevation points 242m, 187m, 184m, 154m and 119m, to elevation point 78m; from this point, descending in a straight line, towards the junction of Estrada do Magarça with Rua Campo Formoso; along this (excluded) to Estrada do Mato Alto; along this (excluded) to the Rio Cabuçu; along its bed, to the junction with Avenida Alhambra; along this (excluded) to Rua Jorge Sampaio; along this (excluded) to Estrada da Cachamorra; following this (included) to Rua Jose Capanema; following this (included) to its end; from there, along the slope, to the highest point of Morro do Cabuçu (elevation 568m); from this point, descending and ascending the slopes and crossing Estrada das Tachas in a straight line, to the highest point of Morro dos Caboclos (elevation 688); from this point, following the ridge in a southerly direction, passing the highest point of Morro da Toca Grande, descending and ascending the spikes and crossing Estrada da Toca Grande at elevation 346m, to elevation 444m; from this point, descending and ascending the spikes, crossing Estrada do Morgado at elevation 248m, to Pico do Morgado (elevation 398m); from this point, along the ridge past Morro da Boa Vista (elevation 334m), to Morro da Ilha (elevation 434); from this point, straight down the slope, until you reach the junction of Estrada da Grota Funda and Avenida das Américas; along this (including only the even side) to Estrada da Barra da Guaratiba; along this (excluded) to the access bridge to the Marambaia Proving Ground (boundary of the ZE-6, Decree No. 4528, of 27/11/70); along this (including and including the stretch of Restinga da Marambaia belonging to the Municipality of Rio de Janeiro) and along the shore of Sepetiba Bay to the starting point, including under its jurisdiction the islands of Guaraguessaba and Urubu and also the islands of Barra de Guaratiba.

152 - **Barra de Guaratiba**

From the Atlantic Ocean, at the place called Saco dos Meros (including and excluding Ponta da Praia Funda), up and down the spur, passing the 96m elevation point, to Garganta da Praia Funda (elevation 13m); from this point, up the spur of Morro de Guaratiba to the 334m elevation point and, along the ridge, to Morro de Sào Joào da Mantiqueira (elevation 273m); from this point, down and up the spur in a northerly direction, passing through Garganta do Grumari (elevation 108m), crossing Estrada do Grumari, to elevation 237m on Morro da Faxina; from this point, along the ridge in a northerly direction, to elevation 394m; from this point, descending and ascending the spikes in a northeasterly direction, passing through points 298m, 358m, 392m, 339m and 413m, to the highest point of Morro de Santo Antônio da Bica (482m); from this point, along the ridge, passing through elevations 452m, 383m and 324m, to elevation 279m; from this point, down the spur, until it meets the junction of Estrada da Grota Funda and Avenida das Américas; along the latter (including only the odd side) to Estrada da Barra de Guaratiba; along the latter (included) to the access bridge to Campo de Provas da Marambaia (boundary of ZE-6, Decree No. 4528, of 27/11/70 (excluded); from there, along Praia da Barra de Guaratiba (included) and along the seafront to the starting point, including under its jurisdiction the islands of Frade and Rasa de Guaratiba.

153 - **Pedra de Guaratiba**

From Praia da Capela, on Sepetiba Bay, following the extension of Avenida Carlos da Silva Rocha and along it (included) crossing Estrada da Matriz, to the junction with Rua Soldado Prim Canes; from this point, along the extension of the alignment of Avenida Carlos da Silva Rocha, to Rua 25 do PAL 18.529; along this (excluded) to Estrada da Capoeira Grande; along this (excluded) to Estrada do Catruz; along this (included) to Estrada da Pedra; along this (included) to Rua Maestro Deozilio; along this (included) and its extension, to Baia de Sepetiba; from there, along its shore including the beaches of Venda Grande, Pedra and Capela, the starting point.

ANNEX B - Urban Plan drawn up by the Guaratiba Rural Association UAP Working Group for the Guaratiba Island area

To ensure that Ilha de Guaratiba remains the Green Neighborhood it always was, the group has come up with the following suggestions:

1 Maintain existing cultivated areas that are registered with the respective associations, with privileges in IPTU and ITR collection Example: Itaguaí, with mixed areas (Residences next to Commerce and Rural areas).

2 Create Zones of Occupation, with the different gauges. 2.1 Exclusively Agricultural Zones - more than 5000 m². The height in these areas would be single-family houses with a maximum of 2 storeys.

2.2 Exclusively residential zones up to elevation 80. Minimum plots of 500 to 800m². The height in these areas would be 4 stories.

2.3 Mixed Zones - Commercial / Residential / Agricultural. Minimum plots of 2000m². The height in these areas would be 4 storeys.

2.4 Agro-forestry zones at elevations 60 to 100. Minimum plots of more than 10,000m²

2.5 Draw up ecotourism projects including areas of Pedra Branca Park.

2.6 Zones free of any occupation: riverbanks, areas subject to flooding, swamps, streams with at least 5 m of riparian forest.

3 Do not allow heavy or polluting industries

4. Draw up a Pilot Plan with future street layouts dividing up the zoning of the areas

5 Draw up basic sanitation project

6 New, individual buildings. Condominiums and allotments must use biodigesters for sanitation.

The UIP should address ways of regularizing the situation of irregular ownership and construction that are prevalent in the region.

ANNEX C - Letter of Recommendation for the PEU - Ilha de Guaratiba Urban Structuring Project - proposed by AMIG - Association of Residents of Ilha de Guaratiba

Rio de Janeiro, RJ, 08/10/2015.

Office No. 002/15

Ref: Letter of Recommendation for the PEU - Ilha De Guaratiba Urban Structuring Project

The Association of Residents of Ilha de Guaratiba hereby requests that you, Mr.ª , consider the "**Letter of Recommendation for the PEU - Proposal for the Urban Structuring of Ilha de Guaratiba**" proposed here, as follows:

SUMMARY OF THE MINUTES OF THE MEETING OF THE ASSOCIATION OF RESIDENTS OF ILHA DE GUARATIBA ON OCTOBER 3, 2015.

In order to collaborate with Rio's City Hall in view of the news about the preparation of the Guaratiba PEU, the **ASSOCIAÇÃO DE MORADORES DE ILHA DE GUARATIBA**, at its meeting on October 3, 2015, prepared this proposal to be sent to the Mayor.

This proposal was divided into 3 parts.

PART 1ª - A deals with the necessary prior organization of the region, in order to guarantee the urban and environmental infrastructure with a view to receiving the changes in the built space as a result of the implementation of the UIP guidelines.

This part is based on the principles set out in **Article 65 of the 2011 DIRECTOR'S PLAN**, which deals with the **REGIONAL PLAN, an** instrument that defines the guidelines, objectives and actions for a Planning Region. In other words, before the UDP is applied, it is necessary to guarantee the basic urban functions (circulation, work, housing and leisure), as well as the social functions of the city as laid down in the **Organic Law of the Municipality.**

2ª PART - B deals with the items to be considered in the text of the PEU with a focus on the organization of built space, according to the nature of this instrument for controlling land use and occupation.

It should be noted that the following items are specific to the **ILHA DE GUARATIBA** (PI) district and the others relate to the entire district governed by the **PEU DE GUARATIBA** (PG).

A - FIRST PART:

Items to be considered by the **Local Development Plan** based on the **REGIONAL PLAN** provided for in **Article 65** of the **2011 Master Plan** to guarantee the structural basis necessary for the consolidation of the UDP.

The following points are covered:

1 - A new road system (PG)

2 - Preventing mineral extraction, including from groundwater (PG)

3 - Sewage treatment and rainwater drainage, garbage, macro-drainage program, water supply (PG)

4 - Maintaining the concepts relating to the **Area of Special Agricultural Interest** (IP)

5 - Delimitation of the Pedra Branca State Park **Buffer Zone**, (PG)

6 - Delimitation and **protection of the foothills** of the Pedra Branca Massif (PG)

7 - Delimitation of the **Special Area of Environmental Interest (AEIA**) of the Pedra Branca Massif under municipal jurisdiction. (PG)

8 - General urban mobility and accessibility program (PG)

9 - Approval of a plan for public transport using LRVs and "jardineira" buses, on local roads directed towards the BRT lines and the Campo Grande train station. (PG)

10 - Local plan for the **Area of Special Social Interest - AEIS**, with the aim of regularizing irregular occupations and allotments, based on the cadastral survey of these areas. (PG)

11 - Plan for the regularization of allotments registered with the **Nùcleo de Regularização de**

City Hall allotments (PG)

12 - Plan to regularize allotments not registered with the **Nucleus** (PG)

13 - Discrimination of **public land** in a publicly accessible cadastre. (PG)

14 - Use of empty and idle areas to create forests and leisure areas (PI).

15 - Inscription in the tax register of existing roads without municipal inscription: those not recognized by the City Hall located at the accesses to, or within, allotments registered with the Allotment Regularization **Nucleus** and those still under INCRA's tax regime, remnants of the former Rural Zone of the Municipality. (PG)

16 - Review and improvement of public lighting; (PG)

17 - Improving education infrastructure; (PI)

18 - Review of education policy in line with local reality; (PI)

19 - Construction of crèches and improvement of the health center (PI)

20 - Afforestation plan for the region using native species.

21 - Preventing the establishment of heliports (PI)

22 - Landscaping of the right-of-way of municipal roads and avenues (PG).

23 - Installation of an Urban and Social Guidance Office at the headquarters of the XXVI RA.

B- SECOND PART - (Items to organize the land use policy for the text of the UIP)

1 - Exclusive **single-family residential** use (PI)

2 - Main economic activity Agriculture and farming with a focus on **ornamental plants** (PI)

3 - Zoning to protect traditional uses for **urban agriculture** (PG)

4 - Construction of **sheds** only as a complement to agricultural activities. (PI)

5 - Delimitation **of** the **Marginal Protection Zone (MPF**) for surface waters (PG)

6 - No gas stations or banks. (PI)

7 - Maximum height of seven meters with a maximum occupancy rate of 50% (PI).

8 - Minimum plot size 600m2 (PI)

9 - Neighborhood commercial use in an exclusive building (PI)

10- Review and improvement of public lighting (PG)

11- Improving education infrastructure (PI)

12 - Review of education policy in line with local reality (PI)

13 - Building nurseries and improving health centers (PI)

14 - Afforestation plan for the region.

15 - Preventing the establishment of heliports (PI)

16 - Landscaping of the right-of-way of municipal roads and avenues (PG)

FINAL REMARKS:

LEGENDS:

PG = Guaratiba Plan valid for the entire neighborhood

PI = Guaratiba Island Plan valid only for the Guaratiba Island area

________X________

FOUNDATIONS OF URBAN POLICY IN THE MUNICIPALITY'S ORGANIC LAW.

Art. 422= Social functions of the city

Art. 422 - Urban policy, formulated and administered within the scope of the planning process and in line with other municipal policies, shall implement the full fulfillment of the social functions of the City.

§ Paragraph 1 - The social functions of the City include the right of the population to housing, public transport, basic sanitation, drinking water, urban cleaning services, drainage of circulation routes, electricity, piped gas, supply, public lighting, health, education, culture, day care, leisure, slope containment, security and preservation, protection and recovery of environmental and cultural heritage.

Art. 427 - Public authorities shall guarantee the population means of access to all information on urban policy, as a means of controlling the responsibility of their actions:

I - in the master plan;

II - in the budget preparation and execution process;

III - in urban and regional development plans;

IV - in the definition of industrial locations;

V - in infrastructure projects;

VI - access to the updated public land register;

VII - information on the management of public services.

x _____

CHAPTER II: URBAN PLANNING INSTRUMENTS

Section I: Plans, Programs and Projects of the Municipal Administration

Art. 65: The Regional Plan is the instrument that defines the guidelines, objectives and actions for a Planning Region, and will be subject to the provisions of this Complementary Law.

§Paragraph 1. The regional plan shall contain:

I - Regional diagnosis, including Technical Reading and Community Reading;

II - articulation of sectoral public policies as provided for in this Complementary Law ;

III - indication of sectoral plans and programs;

IV - the administration's projects and actions for the area;

V - targets and deadlines for implementing actions;

VI - definition of the form of monitoring;

VII - budget forecast.

§ Paragraph 2. The Strategic Plan, Agenda 21 and other bodies that may be created for this purpose or related to it, may be used as planning and cooperation bodies for intersectoral coordination, as referred to in Paragraph 1, item II of this article.

Art. 66: The Sectorial Plan spatializes public policies, establishes sectorial programs and indicates the articulation of the actions of the Administration's sectorial bodies.

Sole paragraph. The Sector Program deals with the structuring of a set of goals and actions for the implementation of one or more related public policies indicated in this Master Plan.

Art. 67: The Urban Project will be designed to implement the policies, guidelines, plans and programs proposed in this Master Plan, with the following objectives :

I - redevelopment of areas of the city;

II - boosting local economies;

III - implementation or expansion of urban and public transport infrastructures;

IV - restoration of historical heritage;

V - implementation or complementation of policies, guidelines, pianos

programs.

Sole paragraph. The Urban Project may be:

I - at the initiative of the Executive Branch or the Legislative Branch, when its realization is a priority for the collective interest;

II - private initiative, when it is a voluntary proposal by a private agent or group of agents;

III - linked to the implementation of a consortium urban operation."

We are sure that our request will receive your attention, and we would like to thank you sincerely.

Sincerely:

Higor Pereira Neto

President

ANNEX D - Guaratiba Urban Structure - Proposal submitted by the Institute of Architects of Brazil

Summary

Throughout August, September and October 2013, the Institute of Architects of Brazil (IAB/RJ) held a series of technical meetings and seminars to study the urban and socio-environmental framework of Guaratiba.

This study was promoted as a reflection by the architects and urban planners who were members of a commission specially appointed by the IAB RJ with the aim of collaborating with Rio de Janeiro City Hall to find the best way forward for the occupation and development challenges facing the region.

We understand that the present and future of the region point to a strong expectation of providing the region with the means to guarantee environmental conservation in an integrated manner with urban development policies.

Having made a quick assessment of the geomorphological, environmental, land occupation, urban planning, social and economic conditions of the region, the special commission of the IAB RJ established the following as the structuring concepts of its recommendations:

The occupation and development of the region must be in line with a General Urban-Environmental Plan.

As the basis for a General Plan, we consider:

- It is necessary to avoid the expansion of the city, because its growth is incompatible with the universalization of the provision of public services according to contemporary requirements;

- Public investment should prioritize the provision of infrastructure and public services for the region's consolidated urban centers;

- To assist families whose homes are in areas at risk, priority should be given to the occupation of vacant land and areas within consolidated areas already equipped with infrastructure;

- Any new construction or occupation must ensure strict compliance with the environmental requirements laid down for the region's nature reserves;

- The General Plan should enable a method of shared management, involving public administration bodies, residents and civil society, for the planning and implementation of the projects planned for the area.

Delimitation of the Area of Special Environmental Interest of the Guaratiba Region, created by Decree 37.483 of July 31, 2013:

Located on the edge of the city's urban sprawl, this region has large unoccupied areas and sparse urban occupations, a typical expression of the peripheral areas of Brazil's large metropolises.

Its territorial size (13,950 ha) is greater than the sum of the neighborhoods in the Center and South Zone.

According to the 2010 Census, Guaratiba has a population of 110,000 inhabitants, living in around 40,000 homes, which means a gross occupational density of only eight inhabitants per hectare and a population growth of around 26% between 2000 and 2010, a considerable percentage compared to other areas of the city.

The opening of the Grota Funda tunnel and the implementation of the BRT crossing this region and directly linking the neighborhoods of Santa Cruz and Campo Grande with the Baixada de Jacarepaguà point to the need to plan the management of this area, from the point of view of its urban occupation, in the short, medium and long term.

Its peculiar relief and the wide variety of rivers crossing the region, from the Pedra Branca massif to Sepetiba Bay, means that the lowlands suffer periodic flooding caused by the regime of the springs and tidal variations. The recent flooding of the area where a World Youth Day event was to be held - "Campus Fidei - proves this characteristic trend of the region.

Rationale

Thinking of the region as an inseparable part of the urban context of the municipality of Rio de Janeiro and its metropolitan region is a condition for addressing the issue.

The elaboration of a planning process must involve organizations of the region's residents, representative of civil society and the various levels of public power that are directly or indirectly involved in the urban and environmental development of the area.

We don't believe that the creation of urban legislation alone can ensure that the region is properly occupied. A set of short-, medium- and long-term actions is needed to provide the means for the government to define the concepts and criteria for the use and occupation of the area, in a way that is coherent with its natural, environmental and geomorphological conditions.

It is therefore necessary to draw up an action methodology that brings together the various municipal, state and federal departments to draw up comprehensive planning and effective management and control of the area's occupation. It is essential to listen to the various social, cultural and business agents working in the region in order to establish, in a transparent way, the main vectors for planning and occupying the area.

In today's world, planning has taken on the character of managing conflicts and particular interests that must be explained and discussed in a clear and objective manner so that they can be confronted with the general interests of the city and its metropolitan region. It is essential that this debate is linked to an efficient communication strategy to ensure it has the visibility and repercussions desired.

The task of conducting this planning process was to be entrusted to an "Executive Planning and Management Commission for the Guaratiba Region", attached to the Civil House of the Rio de Janeiro City Hall, acting in agreement with the State Government and the Federal Government through their designated representatives. The various government departments and public service companies jointly involved in this process would participate with their respective representatives. The legislative powers, the Public Prosecutor's Office and the representative organizations of civil society would act within the scope of their legal attributions.

Two main axes should be focused on when addressing the issue: environmental aspects and urban planning.

The existence of some urban centers deriving from 1950s allotments and some productive and cultural activities installed in the region should be taken into account. There are also some housing estates occupying areas lacking infrastructure and subject to flooding. These and other similar areas will have to be adapted to the conditions of the local environment as a way of establishing a harmonious coexistence between the principles of sustainability and the socio-economic development of the region.

In order to achieve these objectives, it would be advisable for the City Council to create the "Executive Planning and Management Commission for the Guaratiba Region" with the task of establishing and coordinating the planning process for the region. This group would establish the methodological principles and coordinate the actions developed by the themes responsible for drawing up the region's urban planning guidelines.

Developments

To support the "Executive Planning and Management Commission for the Guaratiba Region", the IAB, once it has been set up, will be able to organize debates with the participation of broad sectors that will address various aspects that will provide the basis for greater recognition of the region. Initially, we suggest the following approaches:

- **GUARATIBA AND ITS INSERTION IN THE URBAN and METROPOLITAN CONTEXT.**

The large size of the territory. The existing historical and archaeological heritage. The prospect of urban expansion. The influence of the Barra da Tijuca urban model. The connection with Campo Grande and Santa Cruz. The Grota Funda tunnel and the BRT. The Master Plan. The preliminary studies for the PEU. Current legislation.

Centralities and peripheries. The influence of the cities that make up the Metropolitan Region. Public and private means of transport. Accessibility and urban connection vectors.

- **the local environment**

Preservation and Conservation Units: management, visibility, maintenance and visitation. Restrictions and possibilities for use. Natural parks: recreation, culture and leisure. Watersheds. Nature and environmental reserves. Natural biomes.

- **urban models and land management**

Possible urban planning models. Similar reference examples. The real estate market. The management of urban space in the short, medium and long term. Possible lines of financing. Public-private partnerships.

The documents produced by the various themes should be the subject of an extensive debate aimed at presenting and discussing the ideas and proposals formulated. The contributions made during the debates should be assessed and incorporated into the reports if they are the subject of consensus decisions. At the end, a publication could be produced containing the research material and proposals for the future development of the area.

Annexes:

- DECREE NO. 37.483 OF JULY 31, 2013 - CREATES THE AREA OF SPECIAL ENVIRONMENTAL INTEREST OF THE GUARATIBA REGION - XXVIRA.
- DECREE NO. 37.460 OF JULY 29, 2013 - DECLARES THE AREA IT MENTIONS TO BE OF SOCIAL INTEREST, FOR PURPOSES OF EXPROPRIATION.
- ANNEX 1 ANNEX 1 ANNEX 1 ANNEX 1

DECREE NO. 37.483 OF JULY 31, 2013 DECREE NO. 37.483 OF JULY 31, 2013 DECREE NO. 37.483 OF JULY 31, 2013 DECREE NO. 37.483 OF JULY 31, 2013

CREATES THE AREA OF SPECIAL ENVIRONMENTAL INTEREST OF THE GUARATIBA REGION CREATES THE AREA OF SPECIAL INTEREST

AREA OF SPECIAL ENVIRONMENTAL INTEREST IN THE REGION OF GUARATIBA CREATES AREA OF SPECIAL ENVIRONMENTAL INTEREST IN THE REGION OF GUARATIBA CREATES AREA OF SPECIAL ENVIRONMENTAL INTEREST IN THE REGION OF GUARATIBA -- XXVI RA. XXVI RA. XXVI RA. XXVI RA.

THE MAYOR OF THE CITY OF RIO DE JANEIRO, in the use of the powers conferred upon him by the legislation in force,

WHEREAS the Organic Law of the Municipality of Rio de Janeiro, in the Chapter "On the Environment", art. 463, lists, among others, that "the following are instruments, means and obligations for which the Public Authority is responsible in order to preserve and control the environment: IX - maintenance and defense of permanent preservation areas, understood as those which, due to their physiographic, geological, hydrological, biological or climatological conditions, form an ecosystem of importance in the natural environment, including: mangroves, estuarine areas and restingas; springs and marginal strips for the protection of surface water; vegetation cover that contributes to the stability of slopes subject to erosion and landslides or to the establishment of dunes; areas that are home to rare, endangered or insufficiently known specimens of flora and fauna, as well as those that serve as a place for species to roost, shelter or reproduce; the following natural assets:The Pedra Branca Massif, Parks, ecological and biological reserves, ecological stations and public forests";

CONSIDERING the location in Guaratiba of the Guaratiba Biological Reserve, the Pedra Branca State Park, the Pedra Branca Environmental Protection Area, the Serra da Capoeira Grande Municipal Natural Park; the Serra da Capoeira Grande and Morro do Silvério Environmental Protection Area; the Brisas Environmental Protection Area and the buffer zone for these units;

CONSIDERING the provisions of article 117 of Complementary Law No. 111 of February 1, 2011, which instituted the Master Plan for Sustainable Urban Development of the Municipality of Rio de Janeiro; which classifies the Restinga da Marambaia as a Site of Relevant Landscape and Environmental Interest; the Guaratiba Biological and Archaeological Reserve, the Pedra Branca Massif; the slopes of the Capoeira Grande and Inhoaiba mountains; the Burle Marx Site;

CONSIDERING the risks that the process of densification in this region, following the implementation of the Transoeste highway and the opening of the Grota Funda tunnel, pose to the maintenance of environmental quality, the urban landscape and the quality of life in the Guaratiba region;

CONSIDERING the environmental fragility of the area and the need to prevent the degradation of its environmental conditions, as well as to guarantee the sustainable development of the region, DECREES: -- Art. 1 Art. 1 Art. 1 Art. 1 The Area of Special Environmental Interest (AEIA) of the Guaratiba Region is hereby created, delimited and described in the form of Annexes I and II of this Decree. -- Art. 2 Art. 2 Art. 2 Art. 2 It is hereby suspended, for a period of 180 (one hundred and eighty) days from the publication of this Decree, under the terms of art. 108, § 2 of Complementary Law No. 111/2011 - Master Plan, the licensing of demolition, construction, addition or modification, renovation, transformation of use, land parceling or opening of a street in the Area of Special

Environmental Interest (AEIA) of Guaratiba.

Sole Paragraph Sole Paragraph Sole Paragraph Sole Paragraph Sole Paragraph - Within the period stipulated in the caput, the Municipal Department of Urban Planning, in conjunction with the Municipal Department of the Environment and the IRPH, shall carry out a study with the aim of determining means of protecting the natural and cultural environment of the AEIA. -- Art. 3 Art. 3 Art. 3 This Decree comes into force on the date of its publication.

Rio de Janeiro, July 31, 2013; 449th anniversary of the founding of the City.

EDUARDO PAES

- Annex 2 Annex 2 Annex 2

DECREE NO. DECREE NO. DECREE NO. 37.460 OF JULY 29, 2013 37.460 OF JULY 29, 2013 37.460 OF JULY 29, 2013 37.460 OF JULY 29, 2013

DECLARES OF SOCIAL INTEREST, FOR EXPROPRIATION PURPOSES, THE AREA IT MENTIONS. DECLARES OF SOCIAL INTEREST, FOR EXPROPRIATION PURPOSES, THE AREA IT MENTIONS. DECLARES OF SOCIAL INTEREST, FOR THE PURPOSES OF EXPROPRIATION, THE AREA IT MENTIONS. DECLARES OF SOCIAL INTEREST, FOR EXPROPRIATION PURPOSES, THE AREA IT MENTIONS.

THE MAYOR OF THE CITY OF RIO DE JANEIRO, in the use of his legal attributions, and

WHEREAS World Youth Day (WYD) will take place from July 23 to 28, 2013, in the city of Rio de Janeiro;

CONSIDERING the importance of World Youth Day for the city of Rio de Janeiro and the need to offer a legacy to the people of Rio de Janeiro;

CONSIDERING the need to record the memory of the city and its population of the World Youth Day - WYD Rio 2013;

CONSIDERING the investments already made by the Catholic Church in the area called "Campus Fidei", where the events with the presence of Pope Francis would be held on July 27 and 28, 2013;

CONSIDERING the need to pay tribute to Pope Francis, who has contributed so much, through his words and examples, to awakening a sense of hope and social justice in the population of Rio de Janeiro;

WHEREAS some localities in Guaratiba are characterized and recognized as areas with poor infrastructure, with part of the population living in irregular housing;

WHEREAS it is the constitutional obligation of the Government to ensure decent housing for all, in the form of Article 6 of the Federal Constitution;

CONSIDERING the provisions of art. 5, point "p", of Decree-Law no. 3.365, of June 21, 1941, with art. 2, point V, of Law no. 4.132, of September 10, 1962, DECREES: Art. 1 Art. 1 Art. 1 Art. 1 The area known as "Campus Fidei", located in the Guaratiba district, as per the attached sketch, is hereby declared to be in the social interest, for the purposes of expropriation.

Sole Paragraph - The expropriation will be carried out by judicial means and the total amount should be deducted from the investments made by the Catholic Church in the area in question for the World Youth Day (WYD). Art. 2 Art. 2 Art. 2 This decree comes into force on the date of its publication.

Rio de Janeiro, July 29, 2013; 449th anniversary of the founding of the City.

ANNEX E - Decree No. 24230 of May 20, 2004 - Disposes of the Baixada de Guaratiba, where Ilha de Guaratiba is located, decreeing the creation of the neighborhood called Recreio de Guaratiba.

THE MAYOR OF THE CITY OF RIO DE JANEIRO, in the use of his legal attributions and, considering the experience with the urban planning of Barra da Tijuca; considering the future economic expansion of the metropolitan region with the Port of Sepetiba as its hub; considering the start of work on the Grota Funda Tunnel in a few months; considering the strategic expectation regarding Restinga da Marambaia; considering the availability of public areas,

DECREES

Art. 1: The Municipal Urban Planning Department - directly and through the IPP - will establish within one hundred and eighty days the Urban Planning of the

Guaratiba lowlands - the flat area adjacent to Grota Funda - taking into account:

1 - The municipality is responsible for planning the infrastructure, public spaces, social services, landscaping and boundaries of the new neighborhood;

2 - The definitions for residential and commercial use are the responsibility of the private sector;

3 - The urban planning legislation that needs to be amended, preparing the terms of the bill to be sent to the City Council.

Art. 2: The region mentioned in the previous article will be called Recreio de Guaratiba.

Art. 3: The Municipal Urban Planning Department may use the public project competition as a methodology for parts of the planning process.

Art. 4 This Decree shall enter into force on the date of its publication.

Rio de Janeiro, May 20, 2004 - 440th anniversary of the founding of the city.

CESAR MAIA

Single Annex

Bitmap 1

Referential drawing only

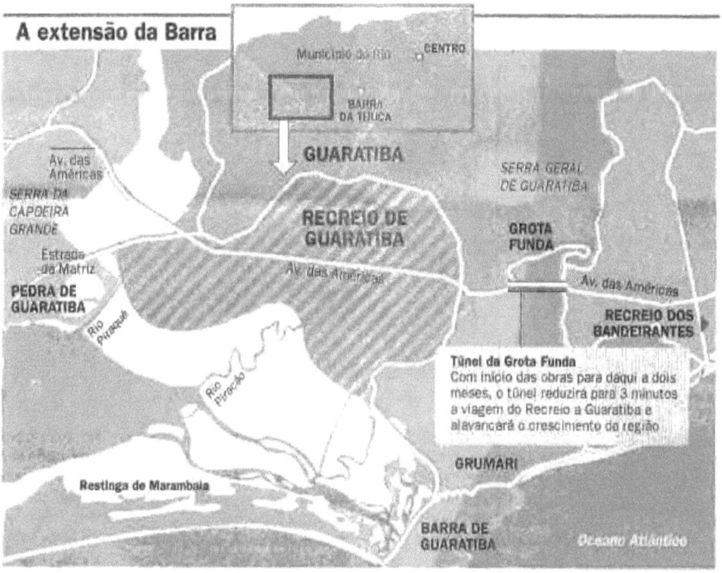
A extensão da Barra
Município do Rio
CENTRO
BARRA DA TIJUCA
GUARATIBA
Av. das Américas
SERRA DA CAPOEIRA GRANDE
RECREIO DE GUARATIBA
Estrada da Matriz
Av. das Américas
SERRA GERAL DE GUARATIBA
GROTA FUNDA
Av. das Américas
PEDRA DE GUARATIBA
RECREIO DOS BANDEIRANTES
Rio Piraquê
Rio Piracão
Túnel da Grota Funda
Com início das obras para daqui a dois meses, o túnel reduzirá para 3 minutos a viagem do Recreio a Guaratiba e alavancará o crescimento da região
GRUMARI
Restinga de Marambaia
BARRA DE GUARATIBA
Oceano Atlântico

MIX
Papier aus verantwortungsvollen Quellen
Paper from responsible sources
FSC® C105338

Printed by Books on Demand GmbH, Norderstedt / Germany